Committee on the Medical Effects of Air Pollutants

Report May 1992 - December 1993

and

Advisory Group on the Medical Aspects of Air Pollution Episodes

Activities Report 1990-1993

LONDON : HMSO

© Crown copyright 1994
Applications for reproduction should be made to HMSO
First published 1994

ISBN 0 11 321881 8
If you require any information about the references used in these reports please write to the Committees' administrative secretariat at:

Department of Health
Room 679D
Skipton House
80 London Road
London SE1 6LW

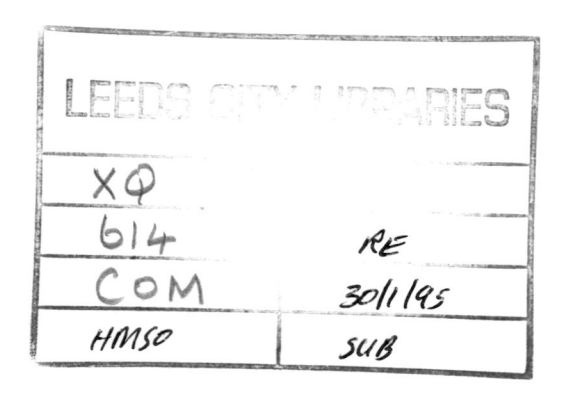

Foreword by the Chief Medical Officer

In keeping with the Government's policy of making the work of Expert Advisory Committees more accessible to the public, we publish the first report of the activities of the Committee on the Medical Effects of Air Pollutants (COMEAP). We also take this opportunity to draw attention to the work of the Advisory Group on the Medical Aspects of Air Pollution Episodes (MAAPE) which has already published three reports.

Members of the Committee and the Advisory Group are independent experts drawn from academia, industry and independent consultancies, and are appointed for a three-year term in the first instance. Membership is reviewed every three years to ensure that it continues to comprise a balanced and appropriate range of medical and scientific expertise. From time to time other experts are also invited to address the Committee and Advisory Group, in the light of advances in relevant science.

I am grateful for the work of the Committee and the Advisory Group and the excellent quality of the advice they have provided. This is highly regarded, not only by the Department of Health, but also by the other Government Departments that seek their opinion.

I look forward to continuing to work with these Groups in the interests of the health of the nation.

K Calman

Contents

Report of Committee on Medical Effects of Air Pollutants (COMEAP)

	Para nos.
Introduction	1
Asthma	2
Benzene	3
Carbon Monoxide	4
Eye irritation	5
Hay fever	6
Health Advice - telephone line	7
Measurement of lung function	8
Monitoring of exposure to air pollutants	9
Open cast coal mining	10
Ozone	11
Particles	12
Sensitive Groups	13
Research	14

Activities report of the Advisory Group on the Medical Aspects of Air Pollution Episodes (MAAPE)

Introduction	1
Ozone	2
Sulphur dioxide	3
Oxides of nitrogen	4
Mixtures of air pollutants	5

Annex A *Membership of COMEAP*
Annex B *Membership of sub-group on asthma*
Annex C *Membership of sub-group on particles*

Committee on the Medical Effects of Air Pollutants (COMEAP)

Preface

Professor S Holgate, (Chairman) MD DSc FRCP

The Committee on the Medical Effects of Air Pollutants was established in May 1992 and forms a part of the Government's continuing commitment to study the effects of environmental factors upon health. The creation of this Committee reflects the importance with which the Department of Health regards air pollution. Its remit is:

At the request of the Department of Health:

(a) To assess, and advise Government on, the effects upon health of air pollutants both in outdoor and indoor air, and to assess the adequacy of the available data and the need for further research.

(b) To co-ordinate with other bodies concerned with the assessment of the effects of exposure to air pollutants and the associated risks to health and to advise on new scientific discoveries relevant to the effects of air pollutants upon health.

The Committee has provided advice on specific questions and has also played a role as a forum for expert discussions and the formation of a consensus view on the health effects of air pollutants. This advice has aided the Expert Panel on Air Quality Standards (established by the Department of the Environment in 1992) in recommending health based ambient air quality standards for the UK. Close links have been maintained with other Department of Health Committees including the Committee on Carcinogenicity.

I have found it stimulating and challenging to take on the job of chairing this Department of Health committee and I am pleased that we have been able to address a number of topics of importance in our first year and a half. I look forward, with enthusiasm, to the Committee's work over the coming years.

Stephen Holgate

Introduction

1.1 This report identifies the main activities of the Committee and the advice which the Committee has provided to Government during the period from its formation in 1992 to the end of 1993. The Committee met five times during that period. The advice provided has been in a number of forms, most often communicated to Government Departments by the Committee's Secretariat following discussion at meetings of the Committee. This advice has been used by the Departments as an input to their work of formulating advice to the public and developing policy.

Asthma

2.1 On 23 October 1992, DH asked the Committee to consider the evidence for trends in the prevalence of asthma, and in mortality from the disease. Members commented on the difficulty of maintaining consistent diagnostic criteria for the disease, and that without these very careful interpretation of the data was required. Changes in diagnostic criteria had occurred over the past 20 years. The Committee's view was that the evidence supported an upward trend, although the rate of the increase was not clear. International comparisons, with asthma increasing in a wide range of countries, meant that it was not at all obvious whether there might be a common factor responsible for the increase. Suggestions made by Members included maternal smoking, changes in diet, and air pollution (exposure to both outdoor and indoor pollutants may be important); genetic factors were also important.

2.2 The meeting agreed to set up a sub-group to consider the relationship of asthma to air pollution further, with the following terms of reference:

To advise on:

(a) Trends in asthma in the United Kingdom

(b) The relationship of air pollution to such trends

(c) Possible mechanisms by which air pollutants might affect trends in asthma

(d) Gaps in the database

(e) Specific research recommendations.

The sub-group was established in 1993 and is expected to report at the end of 1994. Its membership is given in Annex B.

2.3 At the meeting on 22 October 1993, the Committee discussed the report prepared earlier in 1993, for the Medical Research Council's Committee on Toxic Hazards in the Environment and Workplace, by the working group on the Environmental Determinants of Asthma. The Committee noted the importance of distinguishing between the induction of asthma, as opposed to the exacerbation or promotion of an attack. Members felt that research on the former might have to concentrate on the intrauterine environment and early childhood.

Benzene

3.1 On 29 May 1992 the Committee considered the health advice that might be offered to the public on exposure to benzene in the general environment. The Committee concluded that it was long-term exposure to benzene, and therefore long-term average concentrations, that were likely to be of greatest importance in determining the health effects. The use of 1-hour average concentrations was of relevance in traffic and air quality management and other planning measures rather than as a basis for health advice.

3.2 At the meeting of the Committee on 9 July 1993, Members considered the suggestion that benzene might be a significant cause of childhood leukaemia. They noted that no studies in the literature provided any strong and direct support for this hypothesis, and considered that the epidemiology (urban/rural incidence and age profile) and medical evidence (that occupational studies have associated benzene exposure with myeloid leukaemia which is very rare in children, the major childhood leukaemia being lymphoid leukaemia) suggested that it was unlikely to be correct.

3.3 The Committee also considered the need for advice on the health effects of benzene, and what form that advice should take. Members were keen that any advice should be simple to follow and should avoid the use of jargon.

3.4 The Committee provided a statement of advice on the health effects of benzene.

Advice on the health effects of benzene

3.4.1 Benzene is a commonly occurring chemical found in petrol and it is also produced by the combustion of fuel in the engines of motor vehicles. Petrol engine exhausts are the prime source of benzene in atmospheric air although cigarette smoke is a larger source of benzene exposure for smokers.

3.4.2 Long-term occupational exposure to benzene at concentrations a thousand times greater than those typically found in urban air in the UK has been shown to increase the risk of certain types of leukaemia in adults: its use in industry is therefore strictly controlled. Although it is not possible to define an absolutely safe level of exposure to benzene, levels in outdoor air encountered in the UK are considered by COMEAP and the Department's Committee on Carcinogenicity to present a very low risk to health.

3.5 The Expert Panel on Air Quality Standards (EPAQS), taking into account all the available evidence of the health effects of benzene and information on outdoor levels, subsequently recommended to the Department of the Environment an air quality standard of 5 parts per billion (ppb) (running annual average concentration). Since benzene is a genotoxic carcinogen and, in principle, exposure to such substances should be kept as low as practicable, EPAQS also recommended that this standard be reduced to the lower level of 1 ppb in the long term.

Carbon Monoxide

4.1 On 23 October 1992, the Committee discussed the health effects of carbon monoxide. Members concluded that the modelling of these effects was difficult, as the uptake and elimination of carbon monoxide varied with physiological factors. It was agreed that those with ischaemic heart disease were the most sensitive group but that, even in members of this group, exposure to ambient levels of carbon monoxide would be unlikely to give rise to levels of carboxyhaemoglobin which would produce ill effects. In smokers, the ambient exposure would add little to that associated with smoking. It should be recalled that carbon monoxide equilibrates across the lung and therefore at ambient concentrations a heavy smoker will, whilst not smoking, be losing rather than gaining carbon monoxide.

Eye irritation

5.1 At its meeting on 9 July 1993, the Committee heard a presentation from Mr Dennis Swanston on the effects of air pollutants on the human eye. This covered the pharmacological mechanisms involved, and the factors which determine the extent to which damage may be caused to the eye. Defence mechanisms were also considered. In discussion of the presentation the Committee agreed with Mr Swanston that the precise mechanisms by which air pollutants produce eye irritation have not yet been clearly established.

Hay fever

6.1 Also at the meeting on 9 July 1993, the Committee heard a presentation from Dr Penny Fitzharris and Dr Mike Ashmore on the way in which hay fever develops as a disease, and on panel studies during 1990 and 1991 designed to investigate whether air pollutants could modify the relationships between pollen exposure and the severity of a range of symptoms of hay fever. The results suggested that hay fever symptoms might be worse on days of high ozone concentrations, but Dr Ashmore cautioned that a causal relationship with ozone could not be assumed, as other variables might be involved. In particular, meteorological factors (temperature and wind speed) might be important.

Health Advice - telephone line

7.1 At its meeting on 22 October 1993, the Committee was asked to contribute to an evaluation of the Department of the Environment Air Quality Information Line (0800-556677) which provides a recorded message of information and advice on levels of air pollutants and their effects on health. The Helpline was set up in October 1990 and provides, without charge, information of levels of air pollutants and advice on the effects on health of exposure to such levels. Advice on how exposure might be limited is also provided. The text of the health advice provided on the Helpline is given below. The Committee commented on the need to balance a proper appreciation of the complexities and uncertainties with the need for clear advice that people could follow, and endorsed steps to increase public awareness of the issues. They asked the Secretariat to pursue these points with the Department of the Environment, for possible discussion at a future meeting.

7.2 Current text of health advice provided on Air Quality Helpline (September 1994):

> "The following advice on health applies only on days when air quality is poor or very poor. If you have a breathing problem, such as asthma or bronchitis, you may feel a little uncomfortable on these days. You might cough and feel some pain when breathing deeply. It might help to increase your treatment temporarily - you should talk to your doctor about your options.
>
> One general way you can help reduce the effects is to avoid strenuous or outdoor activity when air quality is poor. Also, if children suffer from asthma, they may be more comfortable not taking part in games when air quality is poor; but there is no need to keep them away from school."

Measurement of lung function

8.1 On 9 July 1993, the Committee was asked for advice on the interpretation of the results of the various tests of lung function used to determine the effects of air pollutants upon the respiratory system. The Committee was also asked whether any one test could be recommended in preference to all others for use in this area. Having considered a number of possible approaches - including FEV_1, Peak Flow, and specific airway conductance measurements - the Committee concluded that no single method could be recommended in all circumstances. Peak Flow and FEV_1 measurements were relatively easy and repeatable, and thus could be the measurements most suitable for a range of studies, but more sophisticated measurements might be more appropriate in some cases, particularly when small changes in distal airway function were to be studied.

8.2   Members noted that changes in lung function were likely to be of greater significance among asthmatics and others with impaired lung function, than for those not suffering from such disease. They felt that, in interpreting studies, changes in lung function should be considered in that context. Transient changes were clearly different from permanent changes due to lung damage but, for the former, changes of, for example, 5% in individuals, were not likely to be of health significance. The importance of a shift in average (population) levels of lung function was also considered. It was recognised that, although such shifts might be small, important effects could occur at the lower end of the distribution curve and the effect on individuals with impaired lung function would be proportionally greater.

Monitoring of exposure to air pollutants

9.1   On 23 October 1992, the Committee discussed the monitoring of exposure of populations "at risk" of adverse health effects. Members commented that arrangements for monitoring and assessing of exposure to air pollution in the UK were inadequate in terms of accurate prediction of effects upon health. In particular, they suggested that there was a need for the results of local authority monitoring and DoE monitoring to be brought together.

9.2   On 12 February 1993, the Committee was invited to comment on implications of the work of the Quality of Urban Air Review Group (QUARG), which had published its first report *Urban Air Quality in the United Kingdom* in January 1993. Members noted that QUARG had identified the need for better coordination of local authority monitoring.

Open cast coal mining

10.1   The Committee has considered the suggestion that open-cast coal mining could have a deleterious effect on the health of those living close to the workings. Following discussion at its meetings on 23 October 1992 and 12 February 1993, the Committee agreed a statement of advice on a study on open-cast coal mining.

Advice on Open-Cast Mining

10.1.1   It was accepted by the Committee that little work had been done on the effects of open-cast coal mining upon the health of local residents, especially those with respiratory disorders which might make them more susceptible to the effects of inhaled particles.

10.1.2   A study by Temple (1992) was considered and it was agreed that the results reported were consistent with an association between the opening of an open-cast mine and an increase in asthma consultations. However, the Committee considered that local awareness of the opening of the mine and concern about its possible effects upon health were plausible explanations of the association and that further work would be needed to sustain a causal hypothesis.

Ozone

11.1 On 23 October 1992, the Committee discussed the health advice that might be offered regarding exposure to ozone. The Committee based its views on the report on ozone by the Advisory Group on the Medical Aspects of Air Pollution Episodes (MAAPE) published in 1991. The Committee recognised that developments had occurred in ozone toxicology and made use of updating papers prepared by the Secretariat. Members considered the more important health effects that result after shorter and longer periods of exposure, and the relative merits of expressing any action levels as 1-hour or 8-hour averages. They asked the Secretariat to consider the consistency of advice, in particular in relation to the work of the Expert Panel on Air Quality Standards (EPAQS).

11.2 The Committee further reviewed the position on health advice at its meeting on 22 October 1993 and agreed a statement of advice on the health effects of ozone which was provided to the Department of Health. This was prepared on the basis of advance information on the EPAQS recommendation on an air quality standard for ozone. EPAQS subsequently recommended to the Department of the Environment an air quality standard for ozone in the UK of 50 parts per billion as a running 8-hour average.

Advice on health effects of ozone

11.2.1 Ozone is a powerful oxidising agent produced by photochemical reactions from primary pollutants emitted by motor vehicles and industrial sources. In high concentrations it may produce lung damage though, at levels likely to be encountered in the UK, long-term effects are unlikely to occur. Levels in summer may produce transient respiratory effects in sensitive individuals taking exercise out of doors.

11.2.2 Ozone concentrations are monitored continuously and provided as hourly averages; these will be averaged over rolling eight-hour periods. The standard is likely to be breached on warm summer days when ozone levels rise. The value recommended for the standard is lower than that adopted in the recent EC Directive on ozone. The purpose of the eight-hour standard is to protect health and to encourage reduction of the output of chemicals which contribute to ozone formation. Information already provided on the basis of one-hour average concentrations will continue to be provided on the free telephone Helpline (0800 556677) to allow individuals to take action to avoid or reduce effects.

Particles

12.1 On 23 October 1992, the Committee discussed the health effects of exposure to small particles. Members agreed that the published studies of mortality effects were consistent in demonstrating an association between low levels of particulate air pollution and mortality, but noted that no convincing evidence of a causal relationship had been produced. Members particularly noted that meteorological factors and measures of pollutants could not be assumed to be independent of each other, nor could their effects.

12.2 Members' opinions differed on the significance of the published studies which had examined the effects of low levels of particles on morbidity and indices of lung function. Some felt that, while the associations found were weak, there was some evidence to support the assertion that effects occurred at levels below existing air quality standards. Others felt that the data were unconvincing and no such conclusion could be drawn.

12.3 The Committee also discussed the most appropriate measure of particulate pollution. They did not conclude that any one measure had been clearly established to be the best, but emphasised that in all published work the method of measurement should be carefully specified.

12.4 On 22 October 1993, the Committee agreed to form a sub-group on particulate matter which would prepare a report for consideration by the Committee. The membership of the sub-group is given at Annex C.

Its terms of reference are to advise on:

(a) The current state of knowledge of effects of variations in mass concentrations of suspended particles upon health (excluding occupational exposures).

(b) The value of the measure of particle levels used by the Expanded Urban Network monitoring sites (PM10) as an index or indicator of levels of airborne particles of significance to health.

(c) Gaps in current understanding and the need for future research.

This sub-group was established in 1993 and is expected to report in early 1995.

12.5 At its meeting on 22 October 1993, the Committee considered the chemical composition of fine airborne particles in the UK. They noted that both sampling and analytical techniques differed between reported studies, so that it was never easy to compare the results. Few if any studies commented on the presence of organic matter in samples although it was likely that many particles were of heterogeneous composition. Fungal spores might also be present as a major component.

12.6 At the same meeting, the Committee considered the latest reports on the links between levels of fine particulate matter and health effects. The Committee considered that, while much of the published material showed relationships, it was important to consider all the evidence in a way that some commentators had failed to do. Adequate control of confounding factors is the main problem in interpreting some studies. Without a comprehensive survey, the conclusions could be misleading.

12.7 At the same meeting, the Committee considered what health advice should be given on the effects of fine particles. Members noted that the sub-group on particles was still considering the issue but, at the request of the Department of Health, provided a statement of advice (given below). This indicates how the Committee felt that Departments should approach the issue of control of the levels of fine particles and its relative importance as against other airborne pollutants. In doing so, the Committee recognised that the issues surrounding fine particles are complex, and that Members may need to review the position when the sub-group has finished its work.

Advice on particulate pollution

12.7.1 DH has asked for advice from COMEAP on this issue and the Committee is awaiting a report from its sub-group on particles before providing this advice. However, in the interim, there appears to be sufficient evidence from studies conducted in a number of countries to give cause for concern about the possible effects of current levels of fine particles upon health. This is a complex area and it is not yet clear what role changes in low levels of particles play vis-a-vis changes in levels of other pollutants and temperature and humidity in causing the changes in indices of morbidity and mortality recorded in a number of studies. Despite these difficulties, which will be addressed in detail by the COMEAP sub-group on particles, reductions in levels of fine particles should be welcomed, such an approach being in line with the precautionary approach adopted by the Government on questions of possible effects of a range of toxic substances.

Sensitive Groups

13.1   At its meeting on 12 February 1993, the Committee considered what specific health advice was required for those members of the public who are in some way particularly sensitive to air pollutants. Members commented that there needed to be a clear division between the identification of high risk groups, which might be important in formulating general public health advice, and the specific advice for those who might be affected by air pollutants. For the latter, the effects on the individuals targeted would need to be considered carefully. It would be important that individuals should be in a position to take steps which would prevent a significant effect upon their health and that the curtailment of activity produced should not be counter-productive. This was considered to be especially important with regard to asthmatic children.

13.2   The Committee noted that the issue of advice relevant to babies and young children would need further consideration.

Research

14.1   The Committee first considered research priorities at its meeting on 23 October 1992. Members commented on the importance of maintaining a balance between laboratory studies and epidemiological work, particularly since existing work showed some inconsistencies between results from the two types of study. Also important was the need to assess the magnitude of public health outcomes in any work. The Committee also provided more detailed comments on possible research areas suggested by the Department of Health.

14.2   In discussing indices of lung function on 9 July 1993, the Committee commented that the development of new methods for such measurements would assist future research studies.

14.3   At its meeting on 22 October 1993, the Committee discussed DNA adducts and their potential use as a marker for exposure to air pollutants. It concluded that this technique was potentially of considerable importance, but more work would be needed to develop the technique before its applicability was clear.

Advisory Group on the Medical Aspects of Air Pollution Episodes (MAAPE)

Preface

Professor A Tattersfield MD FRCP

The Advisory Group on the Medical Aspects of Air Pollution Episodes was established in September 1990 to examine the likely health effects of exposure to such episodes of elevated levels of air pollutants as occur in the UK and the need for advice.

Professor Stephen Holgate was the first Chairman of the Group and led it during the production of the reports on Ozone and on Sulphur Dioxide, Acid Aerosols and Particulates. I took over the Chairmanship in June 1992 and since then we have produced a further report on the Oxides of Nitrogen, published by HMSO in December 1993. The conclusions and recommendations of these reports are summarised briefly in the following pages. A further report on the interactive effects of air pollutants is in preparation.

Chairing the Advisory Group has been an exciting and challenging task. All the Members have contributed extensively to the published reports and it has been gratifying to lead a group with such disparate yet focused expertise.

Anne Tattersfield

Introduction

1.1 The Advisory Group on Medical Aspects of Air Pollution Episodes (MAAPE) was established in 1990 with the specific terms of reference:

To consider whether advice about personal protective measures during air pollution episodes should be given by Central Government and, if so, what that advice should be, to whom it should be addressed, and the criteria which should be adopted for the issuing of any advice.

The Advisory Group was first asked to consider these questions with regard to the episodes of elevated ozone concentrations which have occurred in the UK during hot summer weather before progressing to consider sulphur dioxide and nitrogen dioxide.

Ozone

2.1 The Advisory Group's conclusions on ozone were made public in May 1991 with the publication of the full report in August 1991. The conclusions of the Group were brought to the attention of doctors in England by means of a letter from the Chief Medical Officer (CMO).

2.2 The main conclusion of the report was that changes in lung function may occur in people during such episodes of elevated ozone concentration as are found in parts of the UK during periods of hot summer weather. These changes are unlikely to produce irreversible lung damage though individuals who are sensitive to ozone may experience respiratory symptoms, including cough and discomfort on deep inspiration, whilst taking vigorous exercise out of doors. Although individuals with asthma or other respiratory disorders appear to be no more likely than healthy individuals to be sensitive to ozone, the effects of ozone may be more troublesome in individuals who already have some impairment of lung function.

2.3 The group advised that during episodes of elevated ozone concentration, a reduction in outdoor exercise during the latter part of the afternoon would ameliorate the effects in those who are sensitive.

2.4 The Advisory Group recommended that advice should be made available by means of:

The Meteorological Office incorporating information on ozone levels in its weather forecasts when peak hourly concentrations in excess of 100 ppb are anticipated.

The provision of information on the effects of ambient levels of ozone as part of the recorded message currently available on the Air Quality Helpline set up by DoE and the Meteorological Office, so that those who are particularly sensitive to ozone will be in a position to take steps to reduce exposure.

2.5 The Advisory Group made a number of recommendations for further research particularly on interactions between ozone and other pollutants and on the effects of ozone on patients with respiratory problems.

Sulphur dioxide

3.1 The Advisory Group then considered sulphur dioxide, acid aerosols and particulates as its second task and this report was published in October 1992. The conclusions of the Group were again brought to the attention of doctors in England by means of a letter from the CMO.

3.2 The main conclusion of the report was that individuals who do not suffer from respiratory disease will not be affected by such episodes of elevated concentrations of sulphur dioxide as occur in the UK. However asthmatic patients are more sensitive to sulphur dioxide and, in some parts of the UK, levels of sulphur dioxide regularly exceed those at which effects of clinical significance, including tightness of the chest, coughing and wheezing, have been demonstrated in such individuals. These effects are acute and reversible.

3.3 The group recommended that advice should be made available, particularly to those most likely to be affected, when hourly average concentrations of sulphur dioxide are in the range of 125 ppb (357.5 µg/m^3) to 400 ppb (1144 µg/m^3) and further recommended that a warning should be issued when hourly average concentrations of sulphur dioxide exceed or are expected to exceed 400 ppb (1144 µg/m^3). At these levels many asthmatics may experience significant changes in indices of lung function, and symptoms including tightness of the chest, coughing and some breathlessness. These effects may be of sufficient severity to make it advisable for asthmatics to limit exposure, and increase their treatment in consultation, if necessary, with their doctors. Such effects are likely to be mild for the majority of patients and there is no evidence that they have any lasting effect on asthma. Those not suffering from respiratory diseases are not expected to experience any adverse health effects.

3.4 MAAPE recommended that advice should be made available by means of the provision of information on the effects of ambient levels of sulphur dioxide as part of the recorded message currently available on the Air Quality Helpline set up by the Department of Environment (DoE) and the Meteorological Office. It was further recommended that the point of transition from the "Poor" to "Very Poor" band be lowered from 500 ppb (1430 µg/m^3) to 400 ppb (1144 µg/m^3) and that a warning be issued, with revised advice on the Helpline, when Air Quality passes or is expected to pass into the Very Poor band.

3.5 Regarding suspended particulate matter, the Advisory Group concluded that in comparison with conditions in the UK in the 1950s and 1960s, levels of (non-specific) particulate material were low and were not thought to pose a significant threat to health. However, the Advisory Group commented that little monitoring of particulates regarding composition or the mass of particulates per unit volume of air (gravimetric concentration) is undertaken in the UK and thus this conclusion must be regarded as tentative. The situation regarding particulates is now being reviewed by the COMEAP subgroup.

3.6 With respect to acid aerosols the Advisory Group concluded that there were insufficient data available regarding acid aerosol levels, particularly in urban areas, to allow any assessment of likely effects to be made. The Group strongly recommended that methods for monitoring ambient levels of acidity (hydrogen ion concentration) be developed and a basic monitoring network be established.

3.7 MAAPE noted that data were lacking in a number of areas and recommended further research including:

> Research on interactions between sulphur dioxide and related pollutants, and also with other pollutants eg nitrogen dioxide, ozone.
>
> Epidemiological research into the effects of low levels of these pollutants on the prevalence of respiratory disease and on the incidence of asthma attacks.

Oxides of nitrogen

4.1 The Advisory Group's third report covered the oxides of nitrogen and this was published in December 1993. The conclusions of the Group were also highlighted by means of an article in "*CMO's Update*", sent to all doctors in England in 1994.

4.2 The report concluded that the available evidence indicates that individuals not suffering from respiratory disease will be unaffected by such episodes of elevated concentrations of nitrogen dioxide as occur in the UK. When studied in the laboratory there is no consistent difference in sensitivity to nitrogen dioxide between asthmatic patients and normal individuals. However some recent epidemiological studies have indicated that people suffering from respiratory disorders, including asthma, may experience a worsening of their symptoms when ambient levels of nitrogen dioxide and associated pollutants are raised.

4.3 The conclusion that few effects are likely to occur at levels of nitrogen dioxide encountered outdoors in the UK, led the Advisory Group to the recommendation that health warnings and advice regarding nitrogen dioxide episodes should only be issued in exceptional circumstances.

4.4 It was recommended that information on levels of nitrogen dioxide continue to be provided via the telephone Helpline service and that when nitrogen dioxide levels enter the "Very Poor" band (over 300 ppb, 564 µg/m^3) there should be health advice to those suffering from respiratory disorders, including asthma. Only in the unlikely event that the levels exceed 600 ppb (1128 µg/m^3) should there be a warning, possibly accompanied by a press release.

4.5 MAAPE commented that data were found to be lacking in a number of areas and concluded that there was a clear need for further research, including research into the interactions between nitrogen dioxide and other pollutants, such as ozone and sulphur dioxide.

Mixtures of air pollutants

5.1 The Advisory Group is now considering the health effects of interactive combinations of pollutants and is expected to make its fourth and final report in spring 1995.

ANNEX A

Committee on the Medical Effects of Air Pollutants (COMEAP)

MEMBERSHIP *(during the period of this report)*

Chairman
Professor S T Holgate MD DSc FRCP

Members
Professor H R Anderson MD MSc FFPHM
J G Ayres BSc MD FRCP
M L Burr MD FFPHM
R L Carter MA DM DSc FRCPath
Professor B Corrin MD FRCPath
Professor A Dayan MD FRCP FRCPath FFPM CBiol FIBiol
J E Doe BSc PhD
Professor R K Griffiths BSc MB ChB FFCM
Professor R M Harrison PhD DSc CChem FRSC FRMetS FRSH
Professor A J Newman Taylor OBE MSc FRCP FFOM
D Purser BSc PhD
R J Richards BSc PhD DSc
Professor A Seaton MD FRCP FFOM
Professor A E Tattersfield MD FRCP
R E Waller BSc
M L Williams BSc PhD (until December 1992)

Secretariat
R L Maynard BSc MB BCh MRCPath CBiol FIBiol (Medical)
A Wadge BSc PhD CBiol MIBiol (Scientific)
T G Howe BA (Administrative) (until September 1993)
Ms C Douglas (Administrative) (from November 1993)
Miss J Cumberlidge BSc MSc (Minutes)

ANNEX B

COMEAP: Sub-group on Asthma and Air Pollution

MEMBERSHIP

Chairman
Professor H R Anderson MD MSc FFPHM

Members
J G Ayres BSc MD FRCP
Professor A J Newman Taylor OBE MSc FRCP FFOM
D Strachan MSc MD MRCP MFPHM MRCGP
T Tetley BSc PhD
Professor J Warner MD DCH FRCP

ANNEX C

COMEAP: Sub-group on Particles

MEMBERSHIP

Chairman
Mr R E Waller BSc

Members
J G Ayres BSc MD FRCP
Professor R M Harrison BSc PhD DSc FRSC FRMetS FRSH
J Fintan Hurley MA
D Lamb BSc MB BS PhD FRCPath
J McAughey BSc
D G Upshall BSc PhD

ANNEX D

Advisory Group on the Medical Aspects of Air Pollution Episodes (MAAPE)

MEMBERSHIP *(during the period of this report)*

Chairman
Professor S T Holgate (from September 1990 to June 1992)
Professor A E Tattersfield MD FRCP (from June 1992)

Members
Professor H R Anderson MD MSc FFPHM
Dr M Ashmore BSc PLD
Professor S T Holgate MD DSc FRCP
R J Richards BSc PhD DSc
R E Waller BSc
M L Williams BSc PhD (until December 1992)
J S Bower BSc (from January 1993)

Co-opted for Second Report
Professor J G Widdicombe MA DPhil DM FRCP

Co-opted for Third Report
K Miller ChemEng MSc PhD FRCPath

Secretariat
R L Maynard BSc MB BCh MRCPath CBiol FIBiol (Medical)
Mrs K M Cameron BSc MSc (Scientific) (until March 1993)
J B Greig MA DPhil (Scientific) (from April 1993)
T G Howe BA (Administrative) (until September 1993)
Ms C Douglas (Administrative) (from November 1993)
Miss J Cumberlidge BSc MSc (Minutes)

ANNEX E

Committee on the Medical Effects of Air Pollutants
and
Advisory Group on Medical Aspects of Air Pollution Episodes

MEMBERS' INTERESTS *(Declaration of interests during the period of this report)*

*COMMITTEE ON MEDICAL EFFECTS OF AIR POLLUTANTS

†MEMBER OF PARTICLES SUB-GROUP ‡MEMBER OF ASTHMA SUB-GROUP

#ADVISORY GROUP ON MEDICAL ASPECTS OF AIR POLLUTION EPISODES

Member	Personal Company	Personal Interest	Non-personal Company	Non-personal Interest
Prof S T Holgate (Chairman*)#	ICI (Zeneca)	Consultant	Allergene	Consultant
	Fisons	Consultant	CIBA Geigy	
	Upjohn	Consultant	MSD	
	Bayer	Consultant	BOC	
	Pfizer	Consultant	Sandoz	Grant-funded research
	Laboratorios Almiral (Spain)	Consultant	ICI Bayer/Miles	
	Burroughs-Wellcome	Consultant	Glaxo	
	CIBA Geigy	Consultant	Napp Lab.	
Prof A Tattersfield (Chairman#)*	Merck Human Health Division (Merck & Co)	Member of Board of Merck International Advisory Council (MEDAC)	Glaxo Astra Fisons Pharmaceuticals	
	British Gas plc	Shareholder	Genentech Inc Fujisawa Marion Merrell Dow Bayer	Grant-funded research
			Rhone-Poulenc	Occasional consultancy
			Smith Kline Beecham	
			3M Health Care	Occasional consultancy
Prof H R Anderson*#	British Gas plc	Shareholder	NONE	NONE
	London Electricity	Shareholder		
	Thames Water	Shareholder		
Dr M Ashmore#	TBV Science	Consultancy	NONE	NONE
Dr J G Ayres*	Fisons plc	Medical Adviser	EA (Electrical Authority)	Research Fellowship
	Various pharmaceutical companies	Fees for drug studies	BLF/Fisons	Research Grant
			Allen & Hanburys	Research Grant
J Bower#	NONE	NONE	NONE	NONE
Dr M L Burr*	NONE	NONE	Seven Seas Novex Pharma Ltd	Research grant
Dr R L Carter*	Shell plc	Shareholder	NONE	NONE

Member	Personal Company	Interest	Non-personal Company	Interest
Prof B Corrin*	3i	Shareholder	NONE	NONE
	Abbey National	Shareholder		
	British Gas	Shareholder		
	British Airways	Shareholder		
	British Telecom	Shareholder		
	National Power	Shareholder		
	Powergen	Shareholder		
	Rolls Royce	Shareholder		
	Thames Water	Shareholder		
	TSB	Shareholder		
	Vodaphone	Shareholder		
Prof A D Dayan*	Biocompatibles Ltd	Consultant	Glaxo	Research
	Cantab Pharmaceuticals Ltd	Consultant		
	Rhone Poulenc SA	Consultant		
	ML Labs	Consultant		
	SK Beecham Pharmaceuticals	Consultant		
	Sandoz Pharma AG	Consultant		
	British Gas	Shareholder		
	British Petroleum	Shareholder		
	British Steel	Shareholder		
	British Telecom	Shareholder		
	Eurotunnel	Shareholder		
	Glaxo	Shareholder		
	London Electricity	Shareholder		
	National Power	Shareholder		
	Powergen	Shareholder		
	Scottish Power	Shareholder		
	Thames Water	Shareholder		
	TI Group	Shareholder		
Dr J E Doe*	Zeneca Ltd	Employee	ICI plc	Laboratory retained as toxicology consultancy by ICI plc
Prof R K Griffiths*	Hawthorne Consultants	Director	University of Birmingham	Various research grants related to air pollution
Prof R M Harrison*	Associated Octel Company Ltd	Consultant	NONE	NONE
J Fintan Hurley*†	Institute of Occupational Medicine	Employee	Various pharmaceutical companies	Occupational medicine consultancy
Dr D Lamb*†	NONE	NONE	NONE	NONE
Mr J McAughey*†	NONE	NONE	NONE	NONE
Dr Klara Miller#	BIBRA International	Employee	NONE	NONE

Member	Personal Company	Interest	Non-personal Company	Interest
Prof A Newman Taylor*	Proctor and Gamble	Consultant	NONE	NONE
	Glaxo Shell ICI Zeneca Water Company	Shareholder		
	RAF British Airways	Consultant Chest Physician		
Dr D Purser*	Department of Environment - Building Research Establishment	Employee	NONE	NONE
Dr R Richards*#	Xenova Ltd	Research	NONE	NONE
Prof A Seaton*	Soap and Detergent Industry Association	Chairman Medical Advisory Committee	NONE	NONE
	Bayer plc	Occupational Medicine Consultancy		
	Rolls Royce	Shareholder		
	Scottish Power	Shareholder		
	Hydro Electric	Shareholder		
Dr D Strachan‡	Abbey National Building Society	Shareholder	NONE	NONE
Dr T Tetley‡	NONE	NONE	NONE	NONE
Dr D G Upshall†	NONE	NONE	NONE	NONE
R E Waller*#	Seeboard	Shareholder	NONE	NONE
	Scottish Hydro-Electric	Shareholder		
	Scot Power	Shareholder		
Professor J A Warner‡	First London Allergy Clinic	Shareholder	NONE	NONE
	Glaxo Allen & Hanburys Fisons Astra Pharmaceuticals UCB Pharma Gore Vorwerk Miele Powergen	Research Grants to conduct trials of their products		
Professor J G Widdicombe#	NONE	NONE	NONE	NONE
Dr M L Williams*	British Telecom	Shareholder	NONE	NONE

Printed in the United Kingdom for HMSO. Dd. 299997 12/94 C6 9385 1075

Contents

	Page
List of tables	iv
Introduction	1
Background	1
Cancer registration system	1
Acknowledgements	2
Outline of contents	2
Cancer registration in England and Wales	3
The United Kingdom Association of Cancer Registries	8
Cancer registrations, 1992	9
Cumulative risk of cancer	13
Tables	14
Appendix 1 Guidance notes and definitions	66
Cancer	66
Validity	66
Standardised registration ratio (SRR)	68
Populations	68
Regional health authorities	68
Survival	69
Symbols and conventions used	69
Further information	69
Appendix 2 Cancer registries in the United Kingdom: current directors, addresses, telephone and fax numbers	70
Appendix 3 Regional and district health authorities (1992)	72

List of tables

			Page
Table 1	Registrations of newly diagnosed cases of cancer (3rd digit): site, sex and age, 1992.	England and Wales	14
Table 2	Estimated resident population: sex and age, as at 30 June 1992 (figures in thousands).	England and Wales, England, Wales, and regional health authorities	22
Table 3	Rates per 100,000 population of newly diagnosed cases of cancer (3rd digit): site, sex and age, 1992.	England and Wales	24
Table 4	Registrations of newly diagnosed cases of cancer (3rd digit): site, sex and regional health authority of residence, 1992.	England and Wales, England, Wales, and regional health authorities	32
Table 5	Rates per 100,000 population of newly diagnosed cases of cancer (3rd digit): site, sex and regional health authority of residence, 1992.	England and Wales, England, Wales, and regional health authorities	40
Table 6	Standardised registration ratios: site, sex and regional health authority of residence, 1992 (England and Wales = 100).	England, Wales, and regional health authorities	48
Table 7	Registrations of newly diagnosed cases of cancer (4th digit): site, sex and age, 1992.	England and Wales	disk*
Table 8	Rates per 100,000 population of newly diagnosed cases of cancer (4th digit): site, sex and age, 1992.	England and Wales	disk*
Table 9	Cancer mortality to incidence ratios: site, sex and regional health authority of residence, 1992.	England and Wales, England, Wales, and regional health authorities	56
Table 10	Directly age standardised registration rates per 100,000 population: site and sex, 1983 to 1992.	England and Wales	62

* These tables can be obtained on disk directly from ONS, NCRB, Room 2300, Segensworth Road, Titchfield, Fareham PO15 5RR

Introduction

Cancer statistics - registrations 1992 presents data for England and Wales on those patients who were first diagnosed with cancer in 1992 and whose registrations were received at ONS by July 1998.

At the beginning of April 1996, the Office of Population Censuses and Surveys (OPCS) merged with the Central Statistics Office to form the Office for National Statistics (ONS). ONS is responsible for the full range of functions previously carried out by CSO and OPCS, including labour market statistics and registration of births, deaths and marriages. Whilst ONS is responsible for assembling and disseminating UK statistics, no functions held by Scottish or Northern Irish Departments have been transferred to ONS. This volume, therefore, continues to present data for England and Wales only. It is the third to be published following the redevelopment of computer systems at ONS and within a number of regional cancer registries. It updates the provisional figures published in OPCS Monitor MB1 97/1. The contents of tables have been modified (compared with those in reference volumes containing results for 1989 and earlier) in the light of feedback from customers, particular those in the NHS.

Comparable statistics for England and Wales for 1971 to 1991 have been published in the *Cancer statistics - registrations* (Series MB1) reports. For data prior to 1971, statistics have been published in the *Registrar General's Statistical Review of England and Wales, Supplements on Cancer.*

With the completion of the redevelopment of the cancer registration processing system at ONS, the clearing of the backlogs which arose, and the large amount of data enhancement work done to improve the quality of the database, we now anticipate that we will publish the cancer registration statistics for 1993 within 6 months of this volume.

The future publication of provisional results, based on semi-aggregated data supplied to ONS by the regional cancer registries, will depend upon the timeliness of submission of individual records for 1994 onwards by the regional registries to ONS. Provisional results for 1992 were published in ONS Monitor MB1 97/1 in August 1997. Provisional results for registrations up to 1993 (together with mortality results up to 1995) for the four cancers for which Health of the Nation targets[1] have been set - breast, cervix, lung and skin - were published in ONS Monitor MB1 96/2 in November 1996. Estimates of cancer incidence for around 20 major sites for years 1993-1997, based on the corresponding mortality, were published in Monitor MB1 97/2 in July 1998.

Background

Marked changes in the incidence of, and mortality from, cancer have occurred since the beginning of this century. Currently, about one person in three in England and Wales develops a cancer sometime in their life, and cancer now causes about one in four deaths. About 280,000 new cases of cancer are registered every year, and there are about 140,000 deaths from cancer.

It has been estimated that the treatment of cancer accounts for 6 per cent of all NHS hospital expenditure, amounting to over £1 billion a year[2]. Support for research into cancer is currently over £260 million each year; total government expenditure amounts to around £25 million, while spending by charities totals around £125 million and that by the pharmaceutical industry, over £110 million.[3]

Key people involved in cancer prevention and control include scientists investigating the mechanisms which cause cells to become malignant; those carrying out clinical trials to evaluate new treatments; clinicians treating individual patients; public health physicians implementing screening programmes and educating the public; and epidemiologists attempting to characterise high- and low-risk populations, identify causal factors and provide clues to carcinogenic mechanisms.

Evaluation of this work in any coherent way requires a population-based cancer surveillance system which can monitor variations in incidence and survival over time, between places and between different groups in the population.

Cancer registration system

Questions seeking information for the purposes of cancer registration in England and Wales were first asked in the 1920s; a national scheme has been in existence since 1945 - initially centred on the Radium Commission, but from 1947 onwards at the General Register Office, and at its successors, OPCS and, since April 1996, ONS. Complete geographic national coverage was achieved in 1962. Cancer registration is now conducted by ten independent regional registries which collect, on a voluntary basis, data on cancers incident in residents of their regions, and submit a standard data set on these registrations to ONS. In England, each regional health authority (RHA) area in 1992 was covered by its own cancer registry - except that all four Thames RHAs were covered by one registry; the Welsh Office is responsible for the registration of patients resident in Wales (see Appendix 2). A fuller description of the scheme is given below.

Under similar arrangements there is a system of cancer registration in Scotland, co-ordinated by the Information and Statistics Division (ISD) of the NHS in Scotland Common Services Agency in Edinburgh. ISD is a full member of the United Kingdom Association of Cancer Registries (see below). ONS and the regional registries in England and Wales maintain close contacts with ISD and the Northern Ireland registry, and co-operate in several areas, including answering Parliamentary Questions relating to Great Britain or the UK; supplying information for projects such as the preparation of a cancer atlas, and for the examination of clusters of disease by the Small Area Health Statistics Unit at the Imperial College School of Medicine at St Mary's; and assisting the Cancer Research Campaign with information for its UK-based 'Factsheets'.

Acknowledgements

It is with gratitude that ONS acknowledges the work of the regional cancer registries over the years that the national scheme has been in operation, and their close co-operation with the national registry. The full addresses, telephone and fax numbers of the regional registries in England and Wales, and the registries in Scotland and Northern Ireland, are given in Appendix 2. The current directors of the registries are:

Northern & Yorkshire	Professor R Haward (Medical Director)
	Professor D Forman (Director of Information and Research)
Trent	Dr J Botha
East Anglian	Dr T W Davies (General Director)
	Dr C H Brown (Medical Director)
Thames	Ms E Davis (General Manager)
South & West	Dr J A E Smith
Oxford	Dr M Roche
West Midlands	Dr G Lawrence
Merseyside & Cheshire	Dr E M I Williams
North Western	Professor C B J Woodman
Wales	Dr J Steward

Outline of contents

Nearly all of the detailed tables in this and the previous two annual reference volumes[4,5] have been modified in accordance with feedback from customers, after a period of consultation particularly with those in the NHS. **Table 1** contains the numbers of newly diagnosed cases of cancer by site (to the 3rd digit of the ICD9 code), sex and five year age group.

Table 2 presents the 1992 population estimates by sex and five year age group, based on the 1991 census. **Table 3** gives the rates of cancer incidence per 100,000 population by sex and five year age group corresponding to the numbers of cases in Table 1. **Table 4** gives the numbers of cancer registrations and **Table 5** the rates per 100,000 population by sex and regional health authority (RHA as at 1992), and for England and for Wales, separately. **Table 6** gives the standardised registration ratios by RHA by site and sex (England and Wales as base). **Tables 7 and 8** present the numbers and rates per 100,000 population of newly diagnosed cases of cancer, respectively, by site (to the 4th digit of the ICD9 code), sex and age. These tables are not included in this published volume due to their size but are available on disk[6]. **Table 9** contains cancer mortality to incidence ratios by site, sex and regional health authority. **Table 10** gives the directly age standardised rates per 100,000 population of new cancer cases for England and Wales in 1983-1992 (using the European standard population) by site and sex.

The commentary which follows begins with a brief history of the scheme, covering the three reviews of the system published in 1970, 1980 and 1990; the role of ONS; and the setting up of the National Steering Committee on Cancer Registration (now the Advisory Committee). Also described is the establishment of the United Kingdom Association of Cancer Registries. The next sections give the overall results for all cancer sites in 1992, and estimates of the cumulative (lifetime) risk of cancer. Following these are the detailed national and regional tables. Finally, appendices contain guidance notes and definitions and a discussion of some factors relevant to the interpretation of cancer registration data, and information on the cancer registries and regional and district health authorities.

References

[1] *Fit for the future. Second progress report on the Health of the Nation.* Department of Health 1995.

[2] *A policy framework for commissioning cancer services.* A report by the Expert Advisory Group on cancer to the Chief Medical Officers of England and Wales. Department of Health and Welsh Office 1995.

[3] *Expenditure on cancer research 1995/96.* NHS Executive.

[4] ONS. *Cancer statistics - registrations 1990.* Series MB1 no. 23. The Stationery Office (1997).

[5] ONS. *Cancer statistics - registrations 1991.* Series MB1 no.24. The Stationery Office (1997).

[6] Disks are available direct from ONS, NCRB, Segensworth Road, Titchfield, Fareham, Hants, PO15 5RR, telephone: 01329 813759.

Cancer registration in England and Wales

This chapter presents a brief history of the cancer registration system in England and Wales and an outline of the role of the Office for National Statistics (ONS).

Background and early history

Cancer registration is the process of maintaining a systematic collection of data on the occurrence and characteristics of malignant neoplasms and certain non-malignant tumours. The procedure is widely established throughout the world and generally follows guidelines established by bodies such as the International Union Against Cancer (UICC), the International Agency for Research on Cancer (IARC), the International Association of Cancer Registries (IACR), and the World Health Organisation (WHO).

The great and increasing suffering due to cancer was of concern to the Ministry of Health in the early 1920s and with the introduction of radium treatment, a system was initiated in parts of England and Wales to follow the outcome of treated patients. Both the Radium Commission of 1929 and the Cancer Act of 1939 (never implemented because of the war) incorporated the principle that statistical information about cancer patients was essential for planning and operating cancer care services. In 1945, the Radium Commission was designated as the Statistical Bureau to which the data should be sent for final analysis. This work was taken over by the General Register Office in 1947; and the Cancer Act was repealed in 1948 when the National Health Service Act came into force. From that time the General Register Office, its successors OPCS and, more recently, ONS, have collected and processed data forwarded under voluntary arrangements. Since January 1993, it has been mandatory for the NHS, including Trusts, to provide the core items listed in the cancer registration minimum data set to the regional cancer registries; and for the registries to send these data to ONS (see page 6).

The 1960s

In February 1963 a conference was held at the Ministry of Health for the purposes of paving the way for 100% registration of cancer patients and for seeking means of improving the cancer registration scheme. A Working Party agreed on the regional and national objectives of the cancer registration scheme. At the **regional** level, the objectives were to improve the service to the cancer patient through good record keeping and efficient follow-up; and to provide information for local research into the value of treatment and for epidemiological studies, for the planning and assessment of the cancer service, and for the production of national statistics. At the **national** level, the objectives were to produce national statistical analyses likely to assist in the management of the disease and the understanding of it; to cooperate with other Government Departments and outside bodies in any survey aimed at furthering knowledge of the disease; and to participate, by supplying statistical data as required, in the work of international cancer organisations established to carry out research into the cause and course of cancer.

The Working Party spent a considerable amount of time determining what information should be obtained for analysis at the national level, but it was agreed that the information requested should be kept to a minimum - with the intention of obtaining a more complete record and a greater degree of accuracy. The Working Party's report also discussed and agreed recommendations on desirable national and regional tabulations; the elimination of duplicate activity (in data processing); duplicate registrations; dissemination of information; and the unique difficulties of the (then) Metropolitan Regional Hospital Board areas which are now covered by the Thames registry and the South and West Cancer Intelligence Unit based in Winchester (formerly the Wessex registry).

Advisory Committee Report 1970

Following discussions in 1969 between the Department of Health and Social Security (DHSS) and the Registrar General, an Advisory Committee on Cancer Registration was set up. It was requested simply 'to consider and advise on matters of policy and method relating to the national cancer registration scheme', and its members included several eminent epidemiologists in addition to representatives from the DHSS, the registries and (the then) OPCS.

The Committee reviewed the existing scheme, in which each case of cancer was registered first of all on a registration form and the data subsequently transferred onto an abstract card. These were to be updated and resubmitted to OPCS after five, ten and fifteen years. Each registry received, through the machinery of the general system of vital registration and statistics, details of any death in its area where cancer was mentioned on the death certificate (this is known as the 'green card' system after the colour of the paper onto which the death certificate information was copied). Much difficulty had been caused at OPCS by the late submission of abstract cards, and - even worse - of follow-up cards. The quality of data varied considerably among the regions and even the best fell 'rather short' of 100 per cent accuracy in all particulars. The Committee felt that some of the data collected (for example on treatment) were of doubtful value and placed an unnecessary workload on the registries.

There was, however, unanimous agreement that some form of national cancer registration scheme was necessary in order not only to establish national incidence rates and monitor them for purposes of logistic planning and general epidemiological research, but also to permit prospective studies of cancer in selected groups of the population. In addition, information at the international level for comparison with experience in other countries made a valuable contribution to the understanding of the disease.

Revised scheme

A revised scheme was proposed,[1] covering the definition of cases to be registered; the documentation (a revised and shortened abstract card); a nominal index for use by research workers; national tabulations (to be produced by OPCS); and death notifications (green cards). Probably the most important change suggested was that the system of five, ten and fifteen year follow-up abstract cards should be stopped. Instead, cancer registrations would be 'flagged' in the records maintained by the National Health Service Central Register (NHSCR) - another part of OPCS - in Southport, in the same way that deaths were. As non-cancer deaths to persons flagged as cancer-registered could be notified routinely to the registries, this, together with the green cards, would relieve them of the expensive and laborious task of tracing patients clerically (for example by using hospital records or writing to GPs). This revised scheme was introduced in 1971, backdated to cover all registrations whose anniversary date fell on or after 1 January 1971. The essential features of the system (illustrated in Figure A) have now remained unchanged for over 25 years.

Advisory Committee Report 1980

The revised scheme was reviewed some ten years later when the Advisory Committee was reconvened. Its report[2] presented a large amount of national statistics on cancer incidence, survival, prevalence and mortality. It also highlighted the growing demands for information for clinical research; planning, organising and evaluating services for the prevention and treatment of cancer; epidemiological research; and education of the public.

Many of the Committee's comments on areas where problems were being experienced are still relevant today. The Committee re-emphasised the great value of recording the NHS number, and stressed that personal identification data were essential - for the elimination of duplicate notifications, to enable follow-up and calculation of survival rates, and to enable registrations data to be linked (with suitable safeguards) to other data about the same person. They found a substantial degree of variation among the regions in the excess of registrations over deaths; although difficult to interpret, this suggested an equivalent variation in the degree of ascertainment. The report discussed the three main methods of collection: peripatetic staff, hospital staff and the Hospital Activity Analysis (HAA) system. HAA data were often considered to be insufficiently reliable, but the

Figure A The cancer registration system of England and Wales

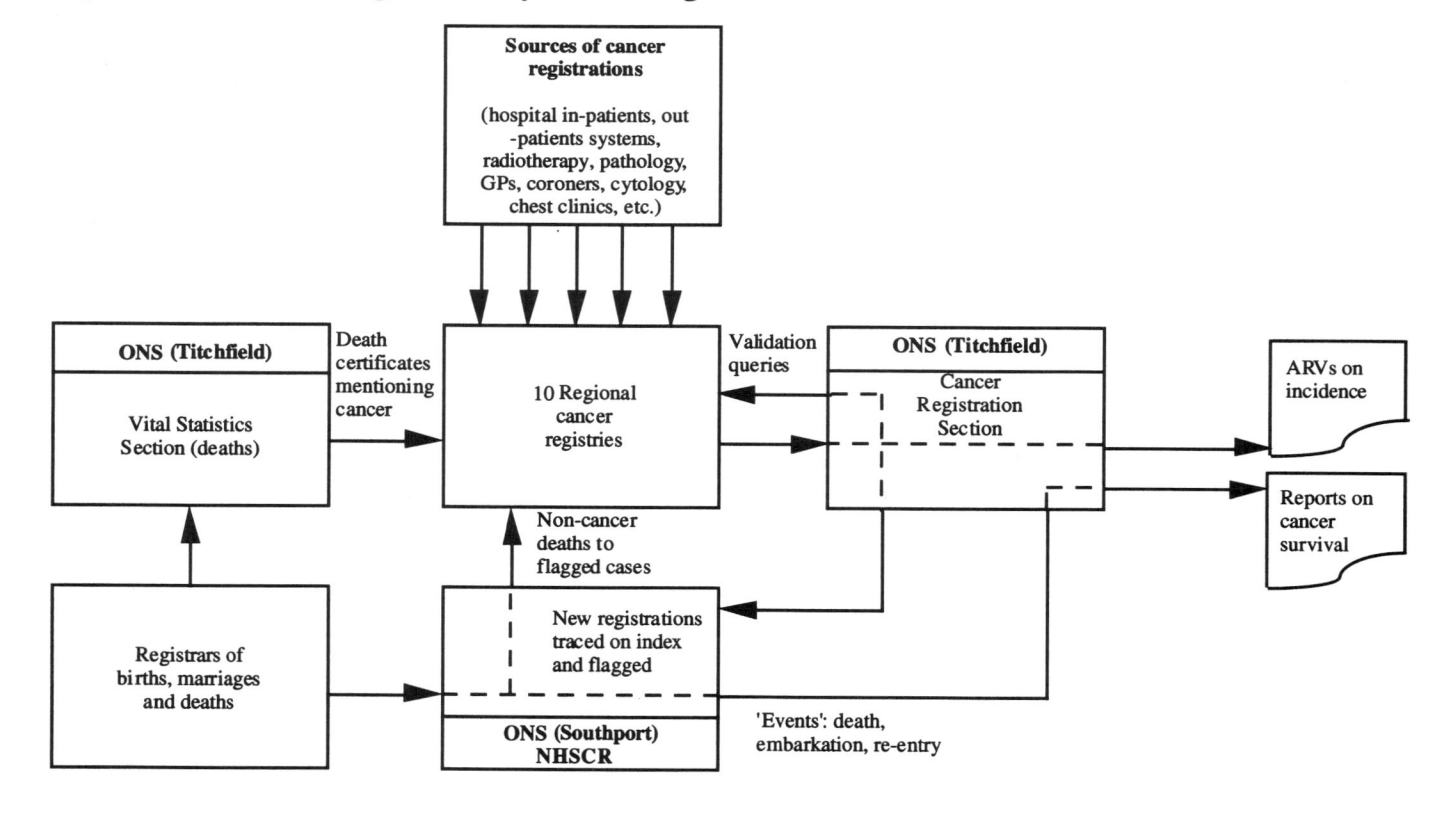

Committee noted that the three regions which used HAA as their primary source were not those which had low numbers of registrations compared with deaths. The use of information from pathology departments, to increase not only the accuracy but also the completeness of ascertainment, was encouraged. As well as being complete, the data needed to be up to date and here the Committee found grave shortcomings since the inception of the revised scheme.

While the average cost of registering one patient with cancer was only a very small fraction of the total cost of the management of the patient's illness, it was noted that (in England) the regional registries were funded by the regional health authorities, with no direct financial input from the DHSS or OPCS. It was possible that registration might not be given the necessary resources at regional level where priorities were decided autonomously.

The Committee concluded that cancer registration covering the whole of England and Wales should continue and be improved in several areas for the following reasons: preventative action was usually based on information from epidemiological studies (using the national register linked to the NHSCR); changes in incidence needed to be monitored because of public, political and medical concern, and improvements in treatment were making mortality data increasingly unreliable as an index of trends; changes in survival needed to be monitored; and reliable and up-to-date data on incidence were essential for the planning and operating of services for cancer detection and treatment.

Medical Advisory Committee review 1990

In 1989 a Working Group of the Registrar General's Medical Advisory Committee (MAC) was set up to review the operation of the cancer registration system, particularly the regional and national data collection methods, the quality and timeliness of the statistics produced, the uses made of the regional and national registers, and the growing tendency to treat cancers in out-patient departments or privately. It was also asked to consider the implications of changes in demand for information and developments in information technology, and the priorities and level of resources required to maintain adequate registers. The potential implications of the recommendations of the White Paper *Working for Patients*[3] were also considered.

The Working Group[4] noted that in addition to the traditional uses of cancer registration (monitoring of time trends and geographical variation in incidence), the system had become vital in several other areas. These included the management of the substantial resources required for the preventative, curative and laboratory services for cancer; the planning and evaluation of services, particularly the screening programmes for breast and cervical cancer; the planning and evaluation of clinical management and treatment based on accurate and unbiased survival data and clinical trials; research into causes of cancer, involving case-control studies and the flagging of cohorts at the NHSCR; and information for health education and health promotion for both professionals and the public. Future uses of cancer registration (especially if linked with other databases) were identified, including evaluating programmes of care, quality assurance, and relating costs to clinical outcome.

The seventeen recommendations made by the Working Group for improvements to the system fell into several categories, relating to the organisation of the system; the collection, processing, quality, timeliness and completeness of the data; and the safeguarding of the necessary data release in view of the impending NHS changes and the use of the private sector.

One of the six recommendations in the 'organisational' area was that a Steering Committee should be established to oversee national cancer registration, with representation from the registries, OPCS, regional and district health authorities, the UK Co-ordinating Committee of Cancer Research, the Health and Safety Executive and the private health sector. This Steering Committee, which was chaired by Dr J Metters, the Deputy Chief Medical Officer at the Department of Health, held its first meeting in June 1991 and met subsequently at approximately six monthly intervals. This committee has recently been re-formed as the Advisory Committee on Cancer Registration, which is chaired by Dr G Winyard, Medical Director at the NHS Executive.

Three recommendations involved both the registries and OPCS: an expanded national core data set; co-operation with the private health sector; and the establishment of guidelines for the handling and release of data. These have been discussed at several consultative meetings with the registries. Work on three other recommendations, relating to the provision of timely estimates of incidence at the national and regional level, quality control checks and the provision of up-to-date anonymous and summary data, is being carried forward at ONS which has recently completed the redevelopment of its longstanding computer system to a new database environment (see below).

The role of ONS in cancer registration

The Office for National Statistics was formed by the merger of OPCS and CSO in 1996. The Director of ONS, Dr Tim Holt, is also the Registrar General for England and Wales. Beneath the Director are six Group Directors, each responsible for particular business areas; these in turn comprise a number of divisions. In the Census, Population and Health Group, the National Cancer Registration Bureau (part of the Population and Vital Statistics Division in Titchfield) conducts the primary data processing of registries' data. Flagging of cancer registrations is carried out by a section at the NHSCR in Southport. Secondary analysis on cancer is conducted within the Health and Demography Division in London, by a statistician and two researchers, supported by a medical epidemiologist and the Deputy Chief Medical Statistician who is also Professor of Epidemiology and Vital Statistics at the London School of Hygiene and Tropical Medicine.

All the work on processing in Titchfield and flagging at the NHSCR in Southport has, since 1993, been paid for by the Department of Health (DH). A service level agreement (SLA) has been negotiated between DH and ONS. Work on the key targets and outputs established in the relevant ONS divisional business plans and the SLA is monitored continuously. ONS makes formal progress reports to the six-monthly meetings of the Joint (DH and ONS) Management Board - and to the Advisory Committee on Cancer Registration.

The regional cancer registries submit notifications of registered patients to ONS on magnetic media (mostly disk). The following data items are specified in the cancer registration minimum data set[5]:

    Record type
    Registration details (identity number)
    Patient's name
    Patient's previous surname
    Patient's address
    Post code
    Sex
    NHS number
    Marital status
    Incidence date
    Site of primary growth
    Date of death (if dead)
    Type of growth
    Behaviour of growth
    Multiple tumour indicator
    Date of birth
    Basis of diagnosis*
    Death certificate only indicator*
    Side (laterality)*
    Treatment(s) (indicators)*
    Stage†
    Grade†

Optional Data
    Ethnic origin
    Country of birth
    Patient's occupation
    Patient's employment status
    Patient's industry
    Head of household's occupation
    Head of household's employment status
    Head of household's industry
    Registration from screening

*   From incidence year 1993
†   From incidence year 1993; phased introduction - initially only for breast and cervix.

The data are loaded onto the new person-based database (see below) and validated. The extensive checks include the compatibility of the cancer site and the associated histology; these checks are closely based on those promulgated by IARC[6].

Redevelopment of the ONS cancer registration computer system

Beginning in 1990, over 20 of the major computer processing systems at OPCS (as it then was) - including births, deaths, the Longitudinal Study (1% linked sample from the censuses), marriages and divorces - were developed to a modern database environment. The two main objectives of the redevelopment of the cancer registration computer system were to have an effective and efficient processing system; and a person-based database (rather than annual files of tumours). To meet the timetable for introducing the new system, it was necessary to convert the 21 annual tumour files (1971 to 1991 inclusive) to a person-based database before the new system began operation. From among the 4.5 million records, those which were either duplicates or were true multiple primary records for the same person were linked together by a probability matching process based on those successfully operated by the Oxford Record Linkage Study, Statistics Canada, and the Information and Statistics Division (ISD) of the Scottish Health Service. Information on linked registrations was sent to the cancer registries for the deletion or amendment of records as appropriate. The essential structure of the cancer registration system in England and Wales, shown in Figure A on p.4 has remained unchanged; but the identification, and the sending to the regional cancer registries, of the death certificates mentioning cancer and the non-cancer deaths to flagged cases, is now done by the new system in Titchfield. In addition, all validation errors are now returned to the appropriate registry for resolution.

In parallel with the work on the redevelopment of the system at ONS, a very large amount of data enhancement work was completed. This included 13,000 new registrations, amendments and cancellations; amendments to about 40,000 records from the probability matching exercise; 15,000 updates of date of death; 25,000 date of birth and date of death discrepancies; 7,000 no trace indicators added to database; and smaller numbers of trace and event rejects, multiple primary cancer queries from registries, mis-traced Welsh records, "dead" now known to be alive, sex discrepancies, partial or invalid postcodes, and embarks. In addition, 36,000 queries from NHSCR about possible multiple primary cancers have now been dealt with.

The backlog of over 600,000 records which had built up in the registries during the time that the person-based database was being constructed has been successfully processed by the NCRB in Titchfield. Priority for the processing of amendments resulting from validation errors was given to data for incidence years 1990 and 1991. At the same time, the NCRB worked steadily through the remaining problems - some left over from the old computer system, and some new ones. These included amendments to the way the system handled the notifications to the registries of death certificates containing a mention of cancer; corrections to records with duplicate identity numbers; re-numbering of some records for one regional registry; and improvements to postcodes. In addition, the revalidation - to the higher standards embedded in the new system - of all the data which had previously been processed on the old computer system has been carried out and queries sent to the regional registries. The new NHS numbers for flagged cases, together with any dates of death, have been sent from the NHSCR to Titchfield, and sent on to the cancer registries. This information should enable both ONS and the registries to amend records for the "immortals" - cases registered alive but whose death was not linked to the cancer registration.

The backlog of over 600,000 records which had been processed in Titchfield was sent to the NHSCR in Southport once the testing of the module of their new computer system which deals with the flagging of cancer cases had been completed. It was known that about 65,000 of these were for people who had died before 1991 when the computerised index was assembled, and so they would not be on the database at NHSCR. These records were therefore stripped off the Titchfield database and sent separately to Southport on paper. Of the remaining records, which were sent on electronic media, it was expected that about 300,000 would match automatically on the system. It was planned to do the batch runs in order, ie the earliest registrations first, to facilitate the determination of true multiple cancers and duplicates. The flagging of the stockpiled registrations for incidence years 1971 to 1990 was completed in January 1997; and the resulting trace and event (death, embark, re-entry) data have been sent to Titchfield and added to the database. All flagging for 1991 and 1992 has now been completed. Work on the remainder of the data, from incidence year 1993 onwards is continuing. At the same time, ONS will attempt to keep earlier incidence years up to date by processing and flagging any "late" registrations received from the cancer registries.

References

1 OPCS. *Report of the Advisory Committee on Cancer Registration.* OPCS (1970).
2 OPCS. *Report of the Advisory Committee on Cancer Registration, 1980: Cancer registration in the 1980s.* Series MB1 no.6. HMSO (1981).
3 Department of Health. *Working for Patients. The Health Service. Caring for the 1990s.* CM 555. HMSO (1989).
4 OPCS. *A Review of the National Cancer Registration System in England and Wales.* Series MB1 no. 17. HMSO (1990).
5 *Minimum data set for the National Cancer Registration System.* EL(92)95. NHS Management Executive Department of Health.
6 Parkin DM, Chen VW, Ferlay J, Galceran J, Storm HH and Whelan SL. *Comparability and Quality Control in Cancer Registration.* Lyons: International Agency for Research on Cancer, 1994 (IARC Technical Report No. 19).

The United Kingdom Association of Cancer Registries

In the past few years, the regional cancer registration system in the UK has been subject to rapid change. With the development of information technology, the pace of change in registration practice has been quickening, and increasing demands for accurate and timely information being made on the cancer registration system. Changes in the organisation of the health service and in the methods of health care delivery have contributed to an increased interest from various authorities and scientists. There are new uses which can and should be made of registration data, such as medical audit and quality assurance of health care, as well as the routine uses which have been made of these data in the past, such as estimation of incidence and evaluation of survival and mortality.

There was widespread awareness both of the need to improve the quality and completeness of cancer registration data, and of the opportunities to do so through the use of information technology. Together with the increased interest from external bodies in using the data, this led to the creation of several groups bringing together cancer registry staff and personnel from OPCS (as it then was) to discuss and resolve matters of common interest.

The longest standing of these is the *Cancer Registries' Consultative Group* (CRCG) which concerns itself essentially with issues of data collection, including coding and data quality. It has representation from all cancer registries in England and Wales, and its members are for the most part registry managers and others closely involved in the day-to-day business of data collection. The *Cancer Surveillance Group* (CSG) was set up in 1989 to meet a perceived need for a forum bringing together those with an interest in the use of cancer data. It has a loose, open and informal membership and structure. Its members include epidemiologists and statisticians, as well as other registry staff. The *Cancer Registries' Information Technology Group* (CRITG) brings together technical experts from the various registries. Education and training was another area of activity thought to be of such importance that it could justify the establishment of another group. There was, however, no forum which brought together registry directors on a regular basis. There was a danger, therefore, with so many different perspectives and forums in which different points of view could be expressed, that the cancer registries might fail to speak with a united voice when, for example, making representations or giving advice to government. With no coherent framework of organisation, there would be a strong possibility of duplication of effort and inadequate communication between the various groups.

It was therefore proposed that a United Kingdom Association of Cancer Registries be established. Following preliminary meetings at which almost all of the UK registries were represented, the Association was brought into being on 2 April 1992 in Cardiff.

The Association has a federal structure. All affiliated population-based cancer registries in the United Kingdom, ONS, the Information and Statistics Division of the NHS in Scotland and the Northern Ireland Cancer Registry are full members with their representative, usually the Director, having a vote on the Executive Committee. Other registries may become associate (non-voting) members; currently these comprise the Childhood Cancer Research Group in Oxford, the CRC Paediatric and Familial Cancer Research Group in Manchester, and the Republic of Ireland national cancer registration scheme. The Chairs of the CRCG, CSG and CRITG, and of the Education and Training Group established in 1993, are invited, as appropriate, to attend Executive Committee meetings as observers.

The current officers are: Chair - Dr T W Davies, Director of the East Anglian Cancer Registry; Vice-Chair - Dr J A E Smith, Director of the South and West Cancer Intelligence Unit; and Treasurer - Mr P Needham, Deputy Director of the Trent Cancer Registry. It was agreed that ONS was the most appropriate body to provide secretariat facilities; Dr A J Fox, the Chief Medical Statistician has nominated Dr M J Quinn (Director of the NCRB) to be the Association's Executive Secretary.

The Association provides:
- a focus for national initiatives in cancer registration;
- a coherent voice for representation of cancer registries in the United Kingdom;
- a channel for liaison between registries and for agreeing policy on matters connected with cancer registration;
- a framework to facilitate the operation of special interest groups and regional registries;
- a means of stimulating the development of cancer registration, of information procedures and practices, and of research based on cancer registry data.

The Association represents the views of its members to government and other bodies operating at national level on issues concerned with data quality, the definition of information requirements, and the development of health information systems where these have implications for cancer registration, in particular where matters of overall policy are concerned. The Association is represented on the recently re-formed National Advisory Committee on Cancer Registration. The establishment of such close links is very important given the intimate ties many regional registries have with NHS information systems, and the potential importance of cancer registration to NHS functions such as medical audit and contracting.

Cancer registrations, 1992

Interpretation

Care is required in the interpretation of cancer registration statistics, particularly when addressing either trends over time or differences between regions.

Registration of cases of cancer is a dynamic process in the sense that the data files both in the regional cancer registries and at ONS are always open. Cancer records may be amended - for example, the site code may be modified should later, more accurate, information become available. The date of death is added for cases registered when the person was alive. Records may be cancelled, although this is relatively unusual. Also, complete new 'late' registrations may be made after either the regional cancer registry, or ONS, or both, have published what were thought at the time to be virtually complete results for a particular year.

Consequently, the figures for registrations published by a regional cancer registry in its reference volume may be different from those in the corresponding ARV published by ONS in the series MB1, which will generally have been produced at a different (usually later) time. In addition, both sets of published figures will differ again from the numbers of registrations currently on the files. Further differences between regional registry and ONS figures may arise if records which have been rejected by the validation process at ONS have not been corrected by the registry concerned before the corresponding ARV tables are produced.

In the section on 'Validity' in Appendix 1, it is noted that the regional cancer registries probably differ in their levels of completeness of registration. It may be difficult to interpret any apparent trends in cancer registrations because the registries are continually striving to increase their levels of ascertainment of cases. Any particularly large increases from year to year in the numbers of registrations for an individual registry are most likely to have arisen because of this.

Other aspects of the cancer registration system which are relevant to the interpretation of the data include: accuracy; completeness of flagging at NHSCR; geographic coverage; duplicate and multiple registrations; registrations from death certificates; clinical and pathological definitions and diagnoses; changes in coding systems; changes in definition of resident population; and error. These are discussed in Appendix 1.

ONS has been advised both by expert epidemiologists and by members of the Steering Committee on Cancer Registration, that non-melanoma skin cancer (ICD9 173) is greatly under-registered. Registration varies widely depending on a registry's degree of access to out-patient records and general practitioners. This under-registration of non-melanoma skin cancer is not just a problem for the registries of England and Wales. *Cancer Incidence in Five Continents Volume VI* [1] reports that cancer registries in the United States, Australia, and parts of Europe, also collect very limited information on skin cancers. In the commentary which follows, the figures for 'all malignancies' **exclude non-melanoma skin cancer.**

Cancer registrations in England and Wales, 1992

In 1992 there were totals of 133,584 registrations of cases of cancer for males and 157,686 for females. In the 9th revision of the International Classification of Diseases (ICD9), malignant neoplasms are coded 140-208 and benign, in situ, uncertain and unspecified neoplasms are coded 210-239. In 1992, of the total registrations about 5,200 for males and 27,800 for females were non-malignant. Around two thirds of the non-malignant neoplasms for females were carcinoma in situ of the cervix (ICD9 233.1).

In 1992 there was an increase of 4,758 registrations (4.5 per cent) for males and an increase of 4,178 registrations (3.9 per cent) for females, compared with the published figures for 1991 (Table A). Most regions showed either small increases or decreases in registrations compared with 1991. There will inevitably be some "late" registrations for 1992, although the proportion will be lower than for the late 1980s, owing to the delayed publication of this reference volume. For 1991, not all Welsh records had been received at ONS when the extract on which the annual reference volume MB1 no. 24 was based, was taken from the database in September 1997. For 1992, the Welsh registry notified ONS of some deletions after the extract on which this volume was based was taken in July 1998. The total numbers of newly diagnosed cases (excluding non-melanoma cancer were: males 1991 - 6891, 1992 - 7288 (increase of 4%); and females 1991 - 6841, 1992 - 7258 (increase of 6%).

Cancer is predominantly a disease of the elderly. The overall crude **rates** of cancer registrations, 436 per 100,000 population for males and 429 for females, conceal wide differences between the sexes and across the age groups, as illustrated in Figure B. The numbers on which this Figure is based are given in Table 3. Following the small decrease in rates after early childhood, rates increased continuously across the age-range for both males and females. A falling off in the rates for the very elderly (85 years and over) may indicate under-registration;

---

[1] Parkin DM et al (eds). *Cancer incidence in five continents. Volume VI.* IARC Scientific Publication no.120. Lyons 1993.

this does not seem to have occurred. Rates of cancer rose more quickly with age in females than in males; this is reflected in the age distribution described below. In the 40-44 age-group, the rate in females was double that for males. Subsequently, the overall rates rose more rapidly for males and were broadly similar to those for females in the 60-64 age-group. After this, the rates rose much more rapidly for males - they were about 45 per cent higher than those for females in the 65-69 age-group and almost double those aged 80-84.

The age distribution of malignant neoplasms is shown in Figure C. The numbers on which this Figure is based are given in Table 1. Of the total of 221,583 registrations, only 1,141 (0.5 per cent) occurred in children aged under 15; of these, 379 (33 per cent) were leukaemias (ICD9 204-208). The percentages of cancers in the five-year age-groups tended to rise earlier in females than in males, owing largely to the influence of the incidence of cancers of the breast (ICD9 174) and of the cervix (ICD9 180). Cancers in those aged under 45 amounted to just under 6 per cent of the total for males and 9 per cent for females. The peaks in the age distributions for both males and females occurred in the 65-79 age groups.

Table A - Registrations of all malignant neoplasms*
1991 & 1992
Regional health authority of residence

Region	Sex	Registrations		% change
		1991†	1992	
Northern	M	6,568	6,631	1
	F	6,609	6,576	-1
Yorkshire	M	7,488	7,715	3
	F	7,732	7,944	3
Trent	M	9,393	10,097	7
	F	9,425	9,935	5
East Anglian	M	4,298	4,338	1
	F	4,375	4,331	-1
North West Thames	M	6,420	6,788	5
	F	6,988	6,908	-1
North East Thames	M	7,248	7,514	4
	F	7,555	7,494	-1
South East Thames	M	7,364	7,823	6
	F	7,833	8,586	9
South West Thames	M	6,039	6,207	3
	F	6,614	6,673	1
Wessex	M	6,852	7,296	6
	F	7,249	7,730	6
Oxford	M	4,851	4,816	-1
	F	5,019	5,204	4
South Western	M	7,469	7,658	2
	F	7,651	8,067	5
West Midlands	M	10,919	11,346	4
	F	10,859	10,927	1
Mersey	M	5,212	5,083	-3
	F	5,296	5,238	-1
North Western	M	8,309	8,525	3
	F	8,632	8,971	4
Wales	M	6,148	7,499	18#
	F	6,232	7,663	19#
England and Wales	M	104,578	109,336	5
	F	108,069	112,247	4
	Total	212,647	221,583	4

* Excluding non-melanoma skin cancer (ICD9 173)
† As published in Table A in cancer statistics registrations 1991 (series MB1 no.24)
# See page 9 for further details

Figure B   All malignant neoplasms (excluding 173): Incidence rates by age-group, 1992

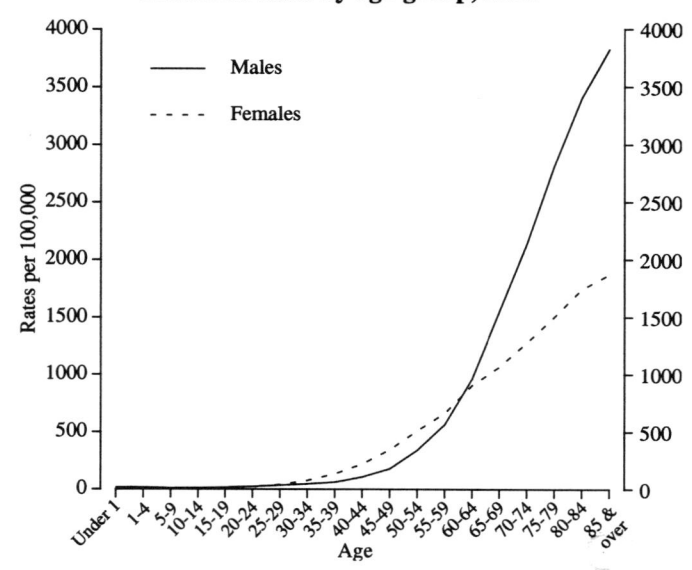

The standardised registration ratios by RHA are illustrated in Figure D. The numbers on which this figure is based are given in Table 6. These SRRs should be interpreted with caution because it is difficult to separate the effect of variation in levels of ascertainment from genuine differences in incidence.

Major cancer sites

In the ICD 9th Revision, there are 62 3-digit site codes relating to malignant neoplasms; of these, four relate to males only and seven to females only. For both males and females just **three** of the sites (different ones for each sex) constituted around half of the total registrations in 1992, as shown in Table B.

The numbers of registrations for the major sites are illustrated in Figure E (and given in Table 1). The numbers of registrations for the 16 sites (counting lip and mouth, colorectal, non-

Figure C   All malignant neoplasms (excluding 173): Frequency distribution by age-group, 1992

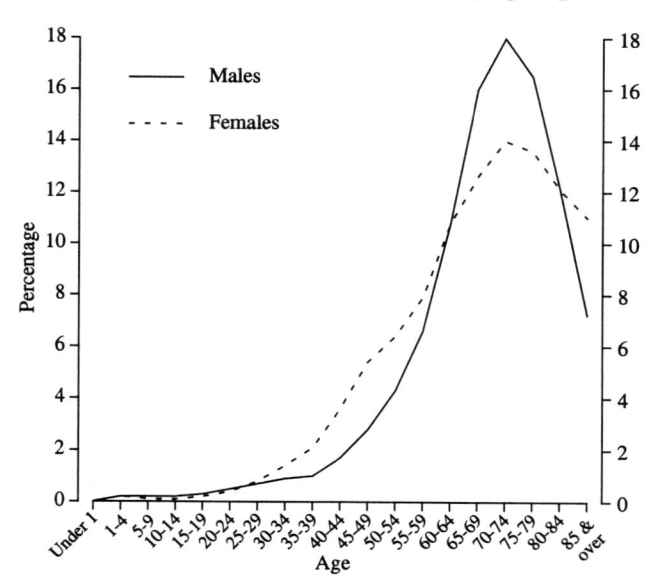

Hodgkin's lymphoma and leukaemia each as one) for males represent 88 per cent of the total; those for the 18 sites for females also represent 88 per cent.

Comparisons with provisional results for 1992

Several years ago, the need was recognised for aggregated regional and national figures for cancer registrations for 1990 to be published before the annual reference volume was produced on the new computer system. Advice was sought from the Steering Committee on Cancer Registration on what data should be collected centrally, bearing in mind particularly the requirements for monitoring the Health of the Nation targets. Based on the Steering Committee's advice and on detailed dicsussions held at a subsequent meeting of the Cancer Registries Consultative Group, the Executive Committee of the UKACR agreed that the registries would supply ONS with the numbers of new registrations in the relevant anniversary year of all malignant sites at the third digit level of the ICD9 code, together with carcinoma in situ of the breast (ICD9 233.0) and cervix (ICD9 233.1), broken down by age and sex (this is essentially similar to the information in Table 1 in this volume).

A pilot exercise was carried out on data for anniversary year 1989 in order to test the procedures both at the registries for producing the information in the required format, and at ONS for receiving and manipulating it; and to highlight any potential major artefacts which could arise. It transpired that there were only marginal differences (of about 1 per cent across most regions and cancer sites) between the data sent directly by the cancer registries and those processed by the National Cancer Registration Bureau. This confirmed the validity of the procedure for the collection of the aggregated data.

Provisional results for 1990 were published in ONS Monitor MB1 95/1 *Registrations of cancer diagnosed in 1990, England and Wales* in July 1995. Comparisons of the results in the annual reference volume for 1990 (Series MB1 no. 23) for all malignancies (excluding non-melanoma skin cancer) with those in Monitor MB1 95/1 indicated differences of only about 100 registrations (0.1 per cent) for both males and females. For the major cancer sites, most of the differences between the two sets of figures were below 1 percentage point.

Provisional results for 1992 were published in ONS Monitor MB1 97/1 *Registrations of cancer diagnosed in 1992, England and Wales* in August 1997. Comparisons of results in this volume with those in Monitor MB1 97/1 indicate differences of just over 1,000 registrations (1.0 per cent) for both males and females (Table C).

Table B   The three most common cancers, 1992

	ICD9	Site description	Number of registrations	% of total malignancies
(a)	**Males**			
1	162	Lung	24,985	23
2	185	Prostate	15,705	14
3	153,154	Colorectal	14,930	14
		Total	55,620	51
		All malignancies*	109,336	100
(b)	**Females**			
1	174	Breast	31,843	28
2	153,154	Colorectal	14,734	13
3	162	Lung	12,327	11
		Total	58,904	52
		All malignancies*	112,247	100

* Excluding non-melanoma (ICD9 173).

Figure D   All malignant neoplasms (excluding 173): Standardised registration ratios by RHA, 1992

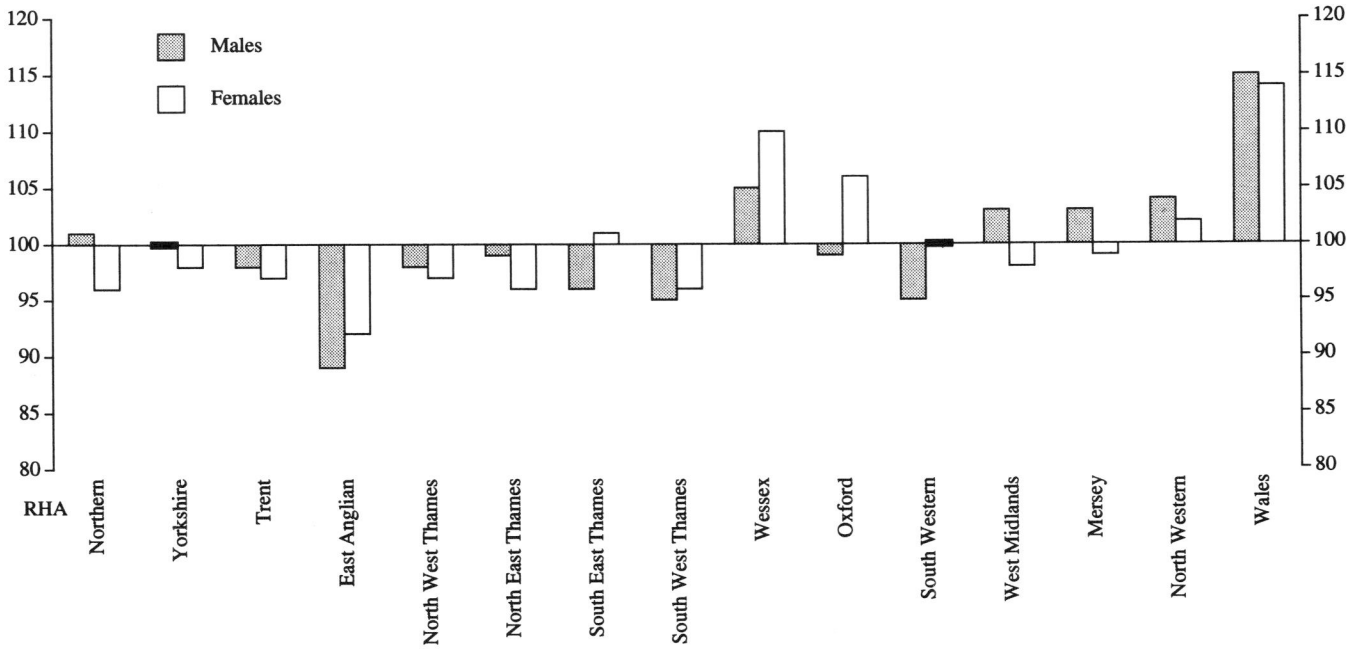

Figure E Registrations - major sites, 1992

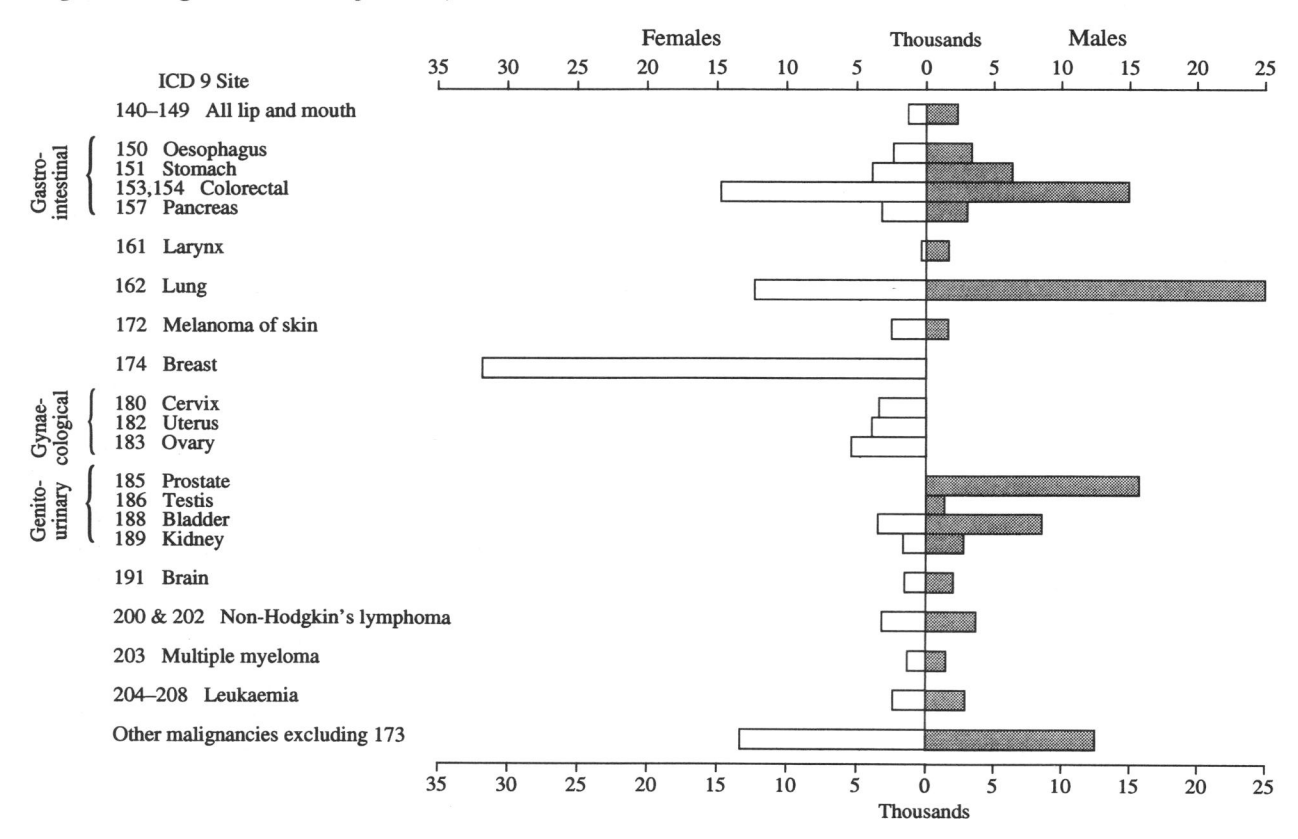

ICD 9 Site	
140–149	All lip and mouth
150	Oesophagus
151	Stomach
153,154	Colorectal
157	Pancreas
161	Larynx
162	Lung
172	Melanoma of skin
174	Breast
180	Cervix
182	Uterus
183	Ovary
185	Prostate
186	Testis
188	Bladder
189	Kidney
191	Brain
200 & 202	Non-Hodgkin's lymphoma
203	Multiple myeloma
204–208	Leukaemia
	Other malignancies excluding 173

Gastro-intestinal: 150, 151, 153/154, 157
Gynaecological: 180, 182, 183
Genito-urinary: 185, 186, 188, 189

ONS is grateful to the directors and staff of the cancer registries in England and Wales for their co-operation in supplying semi-aggregated data during the period when, owing to the redevelopment of the ONS computer processing system, it was not possible to publish national results based on information derived from individual records.

Table C Registrations of malignant neoplasms* 1992: comparison of ARV with provisional results† - main sites

ICD (9th Revision number)	Site description	Males Provisional	Males ARV	% difference	Females Provisional	Females ARV	% difference
140-149	All lip and mouth	2,237	2,317	3	1,296	1,322	2
150	Oesophagus	3,350	3,357	0	2,359	2,380	1
151	Stomach	6,255	6,344	1	3,852	3,885	1
153-154	Colorectal	14,669	14,930	2	14,535	14,734	1
157	Pancreas	3,022	3,025	0	3,176	3,213	1
161	Larynx	1,671	1,681	1	373	365	-2
162	Trachea, bronchus and lung	24,770	24,985	1	12,255	12,327	1
172	Malignant melanoma of skin	1,625	1,650	2	2,439	2,501	2
174	Female breast	-	-	-	31,526	31,843	1
180	Cervix Uteri	-	-	-	3,597	3,400	-6
182	Body of uterus	-	-	-	3,894	3,912	0
183	Ovary and other uterine adnexa	-	-	-	5,272	5,388	2
185	Prostate	15,792	15,705	-1	-	-	-
186	Testis	1,382	1,382	0	-	-	-
188	Bladder	8,528	8,536	0	3,497	3,476	-1
189	Kidney and other and unspecified urinary organs	2,745	2,793	2	1,647	1,671	1
191	Brain	1,967	2,029	3	1,527	1,562	2
200 & 202	Non-Hodgkin's lymphoma	3,665	3,702	1	3,210	3,186	-1
203	Multiple myeloma	1,507	1,502	0	1,364	1,358	0
204-208	All leukaemias	2,887	2,927	1	2,355	2,382	1
	Other malignancies*	12,154	12,471	3	12,922	13,342	3
	All malignant neoplasms* (140-208 x 173)	108,226	109,336	1	111,096	112,247	1

* All malignant neoplasms 140-208 excluding 173   † ONS Monitor MB1 97/1

Cumulative risk of cancer

The cumulative risk of a person being registered with a malignant neoplasm (ICD9 sites 140-208 excluding 173) can be estimated, for males and females separately, by applying sex- and age-specific cancer registration rates to the person years at risk derived from the numbers of survivors from a cohort based on an England and Wales life table[1]. Such a cohort is hypothetical, not a birth cohort, being entirely dependent on the age-specific death rates prevailing in the year for which it was constructed.

For example, for males aged 65-69 there would be 380,740 person years at risk. The cancer registration rate for all malignant neoplasms (excluding ICD9 173) in 1992 for this age-group was 1,552.1 per 100,000 (see Table 3). Thus one would expect there to be

$$380{,}740 \times 1{,}552.1 \div 100{,}000 = 5{,}909 \text{ registrations}$$

or 5.91 per cent of the original cohort.

These age-specific percentages, and the cumulative percentages of risk are illustrated in Figure F. It can be seen that 40 per cent of the cohort of males and 38 per cent of the female cohort would eventually be registered with some form of malignancy. However, registrations would not be equally spread across age-groups. Only 7 per cent of the cohort of males (one sixth of the total) and 10 per cent of the cohort of females (just over one quarter of the total) would be registered at ages below 60.

The cumulative risks of cancer illustrated in Figure F are slightly higher than given in reference volumes containing results for incidence years 1989 and earlier. This is partly due to a methodological modification, ie the use of single years of age, instead of an abridged, life table to calculate person years at risk. Other influences on the cumulative risk are falling overall mortality, and increases in incidence rates - particularly of prostate cancer which is most common in older men.

Reference

1 ONS. *English Life Tables No. 15 1990-1992, England and Wales*. Series DS no. 14. The Stationery Office (1997).

Figure F Cumulative risk of cancer registration

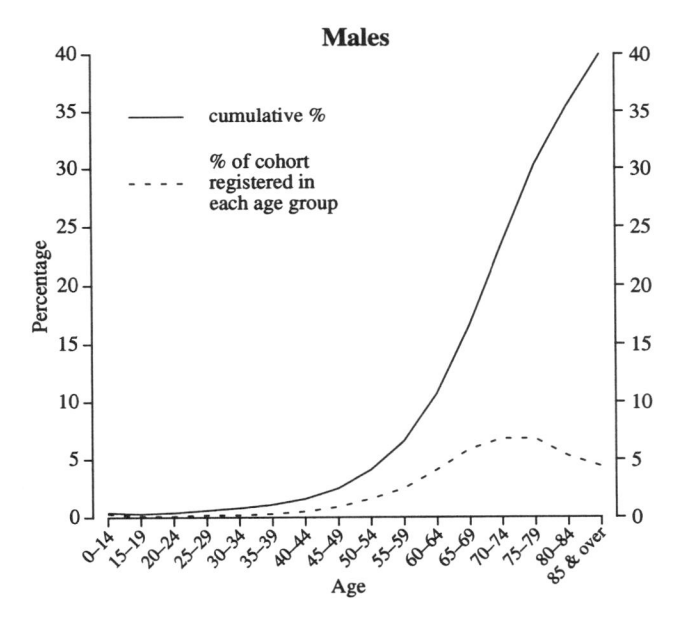

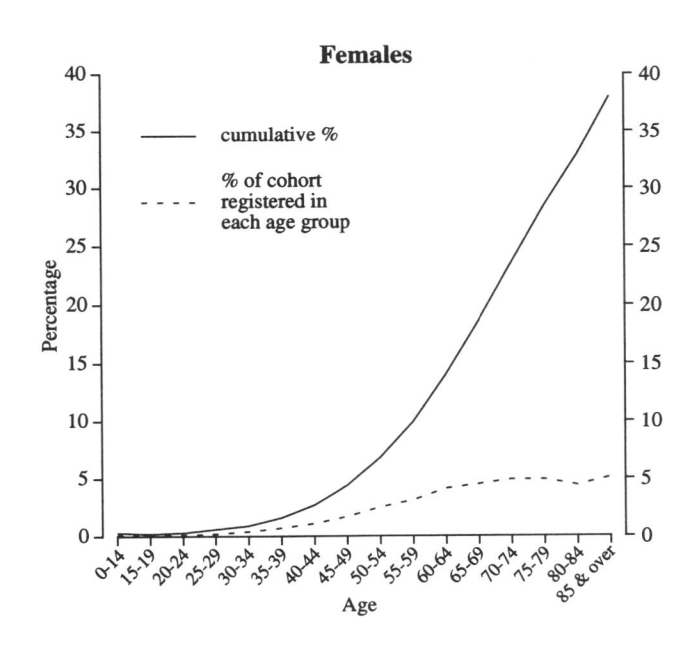

Table 1 Series MB1 no. 25

Table 1 Registrations of newly diagnosed cases of cancer: site, sex and age, 1992

ICD (9th Revision) number	Site description		All ages	Age group								
				Under 1	1-4	5-9	10-14	15-19	20-24	25-29	30-34	35-39
	All registrations	M	133,584	72	263	196	217	334	600	1,005	1,183	1,432
		F	157,686	38	213	152	161	691	3,594	6,024	5,938	5,655
140-208	All malignant neoplasms	M	128,423	67	246	179	200	298	531	889	1,070	1,295
		F	129,905	31	203	136	142	278	525	1,014	1,701	2,603
140-208 x 173	All malignant neoplasms excluding 173	M	109,336	66	245	170	186	277	503	792	946	1,083
		F	112,247	31	198	125	120	243	457	883	1,515	2,309
140-149	Lip, mouth and pharynx	M	2,317	-	-	1	4	4	7	31	22	46
		F	1,322	-	1	3	3	9	7	14	14	24
140	Malignant neoplasm of lip	M	196	-	-	-	-	-	-	1	2	2
		F	63	-	1	1	-	-	-	1	1	1
141	Malignant neoplasm of tongue	M	475	-	-	-	1	2	1	8	5	6
		F	299	-	-	-	-	-	-	5	4	4
142	Malignant neoplasm of major salivary glands	M	236	-	-	-	-	-	1	6	3	8
		F	213	-	-	1	3	3	2	5	4	10
143	Malignant neoplasm of gum	M	69	-	-	-	-	-	1	-	1	1
		F	71	-	-	-	-	-	-	-	-	-
144	Malignant neoplasm of floor of mouth	M	235	-	-	-	-	-	-	-	-	2
		F	99	-	-	-	-	-	-	-	-	-
145	Malignant neoplasm of other and unspecified parts of mouth	M	298	-	-	-	1	1	2	3	7	10
		F	187	-	-	1	-	4	1	-	4	2
146	Malignant neoplasm of oropharynx	M	309	-	-	-	-	-	-	1	-	5
		F	107	-	-	-	-	-	1	1	-	-
147	Malignant neoplasm of nasopharynx	M	138	-	-	1	2	1	2	11	4	10
		F	64	-	-	-	-	-	2	3	2	1	4
148	Malignant neoplasm of hypopharynx	M	258	-	-	-	-	-	-	-	-	1
		F	157	-	-	-	-	-	-	-	-	2
149	Malignant neoplasm of other and ill-defined sites within the lip, oral cavity and pharynx	M	103	-	-	-	-	-	-	1	-	1
		F	62	-	-	-	-	-	-	-	-	1
150	Malignant neoplasm of oesophagus	M	3,357	-	-	-	1	-	-	3	7	15
		F	2,380	-	-	-	-	-	-	2	2	8
151	Malignant neoplasm of stomach	M	6,344	1	1	-	-	-	2	6	13	32
		F	3,885	-	-	-	-	-	3	6	14	18
152	Malignant neoplasm of small intestine, including duodenum	M	249	-	-	-	-	-	-	1	2	5
		F	233	-	-	-	-	-	-	3	3	4
153,154	Colorectal neoplasms	M	14,930	-	-	-	-	3	11	17	52	87
		F	14,734	-	-	1	-	4	7	14	34	78
153	Malignant neoplasm of colon	M	8,573	-	-	-	-	2	8	7	28	50
		F	9,973	-	-	1	-	2	5	11	21	49
154	Malignant neoplasm of rectum, rectosigmoid junction and anus	M	6,357	-	-	-	-	1	3	10	24	37
		F	4,761	-	-	-	-	2	2	3	13	29
155	Malignant neoplasm of liver and intrahepatic bile ducts	M	919	-	1	2	3	1	6	3	4	7
		F	570	-	1	-	2	3	3	5	4	5
156	Malignant neoplasm of gallbladder and extrahepatic bile ducts	M	578	-	-	-	-	-	-	1	2	1
		F	743	-	-	-	-	-	-	-	1	4
157	Malignant neoplasm of pancreas	M	3,025	-	-	-	-	1	-	5	10	21
		F	3,213	-	-	-	-	-	-	2	-	3	16
158	Malignant neoplasm of retroperitoneum and peritoneum	M	115	-	-	-	1	1	2	-	3	-
		F	141	-	-	-	-	-	1	2	2	3

England and Wales
Registered by July 1998

40-44	45-49	50-54	55-59	60-64	65-69	70-74	75-79	80-84	85 and over		Site description	ICD (9th Revision) number
2,468	4,031	5,853	8,986	14,374	21,075	23,780	21,808	16,221	9,686	M	All registrations	
6,562	8,162	8,951	10,784	14,472	16,755	18,740	18,799	16,803	15,192	F		
2,289	3,782	5,582	8,582	13,832	20,321	22,976	21,128	15,723	9,433	M	All malignant neoplasms	140-208
4,346	6,559	7,757	9,583	13,193	15,681	17,769	17,782	16,014	14,588	F		
1,903	3,074	4,670	7,251	11,838	17,527	19,658	17,990	13,277	7,880	M	All malignant neoplasms	140-208
3,890	5,873	6,972	8,598	11,837	13,776	15,284	14,845	13,279	12,012	F	excluding 173	x 173
87	175	212	264	297	377	312	245	145	88	M	Lip, mouth and pharynx	140-149
51	59	87	102	147	174	200	157	115	155	F		
5	6	12	17	18	32	48	28	14	11	M	Malignant neoplasm of lip	140
2	2	3	3	7	5	7	10	10	9	F		
20	45	46	56	56	69	51	68	30	11	M	Malignant neoplasm of tongue	141
17	11	19	31	27	45	51	31	21	33	F		
5	10	16	18	32	43	34	24	22	14	M	Malignant neoplasm of major salivary glands	142
7	6	15	15	19	24	28	27	17	27	F		
-	5	5	5	11	13	8	4	6	9	M	Malignant neoplasm of gum	143
-	2	-	4	11	10	16	11	6	11	F		
6	23	28	39	44	36	28	15	9	5	M	Malignant neoplasm of floor of mouth	144
1	9	13	8	14	17	12	13	7	5	F		
15	24	26	32	34	44	30	31	17	21	M	Malignant neoplasm of other and unspecified parts of mouth	145
3	9	10	12	22	26	26	17	16	34	F		
13	29	41	33	40	60	44	22	15	6	M	Malignant neoplasm of oropharynx	146
11	7	13	8	15	12	15	14	5	5	F		
15	15	10	27	8	12	8	8	2	2	M	Malignant neoplasm of nasopharynx	147
7	6	2	3	7	5	7	2	6	7	F		
5	12	21	24	39	52	41	33	22	8	M	Malignant neoplasm of hypopharynx	148
-	5	11	11	17	26	31	25	19	10	F		
3	6	7	13	15	16	20	12	8	1	M	Malignant neoplasm of other and ill-defined sites within the lip, oral cavity and pharynx	149
3	2	1	7	8	4	7	7	8	14	F		
56	112	172	285	440	582	614	516	351	203	M	Malignant neoplasm of oesophagus	150
12	44	74	111	167	270	380	469	427	414	F		
75	136	230	404	705	1,082	1,164	1,100	870	523	M	Malignant neoplasm of stomach	151
40	57	67	136	244	409	577	731	766	817	F		
12	12	20	21	30	41	39	25	24	17	M	Malignant neoplasm of small intestine, including duodenum	152
4	3	8	11	23	34	40	43	35	22	F		
202	432	734	1,119	1,771	2,456	2,675	2,481	1,809	1,081	M	Colorectal neoplasms	153,154
182	342	511	781	1,292	1,837	2,276	2,474	2,470	2,431	F		
114	200	373	594	970	1,390	1,525	1,499	1,124	689	M	Malignant neoplasm of colon	153
109	220	324	501	861	1,240	1,533	1,649	1,741	1,706	F		
88	232	361	525	801	1,066	1,150	982	685	392	M	Malignant neoplasm of rectum, rectosigmoid junction and anus	154
73	122	187	280	431	597	743	825	729	725	F		
14	18	50	74	111	158	173	138	106	50	M	Malignant neoplasm of liver and intrahepatic bile ducts	155
10	15	13	37	62	69	78	88	79	96	F		
6	6	26	38	71	98	107	94	84	44	M	Malignant neoplasm of gallbladder and extrahepatic bile ducts	156
9	11	27	37	75	84	103	135	131	126	F		
48	70	132	233	360	450	541	539	355	260	M	Malignant neoplasm of pancreas	157
32	52	96	149	247	396	534	582	601	503	F		
5	9	12	18	14	15	16	8	10	1	M	Malignant neoplasm of retroperitoneum and peritoneum	158
4	5	9	6	18	22	24	21	9	15	F		

15

Table 1 Series MB1 no. 25

Table 1 Registrations - *continued*

ICD (9th Revision) number	Site description		All ages	Under 1	1-4	5-9	10-14	15-19	20-24	25-29	30-34	35-39
159	Malignant neoplasm of other and ill-defined sites within the digestive organs and peritoneum	M F	260 315	- -	- -	- -	- -	- -	- -	- 1	1 -	1 3
160	Malignant neoplasm of nasal cavities, middle ear and accessory sinuses	M F	231 185	- -	2 1	- -	- -	- 1	4 2	3 2	2 1	5 3
161	Malignant neoplasm of larynx	M F	1,681 365	- -	- -	- -	- -	- 1	- -	- 1	2 2	10 2
162	Malignant neoplasm of trachea, bronchus and lung	M F	24,985 12,327	2 -	- 2	- -	1 -	3 1	1 4	10 7	27 14	56 46
163	Malignant neoplasm of pleura	M F	921 216	- -	- -	- -	- -	- -	- -	1 -	- -	4 -
164	Malignant neoplasm of thymus, heart and mediastinum	M F	107 69	- -	1 4	1 -	- -	1 1	3 4	4 3	4 -	11 3
165	Other malignant neoplasms within the respiratory system and intrathoracic organs	M F	2 5	- -	- -	- -	- -	- -	- -	- -	- -	- -
170	Malignant neoplasm of bone and articular cartilage	M F	231 167	- -	4 -	5 3	14 15	29 16	19 14	11 8	9 4	8 5
171	Malignant neoplasm of connective and other soft tissue	M F	681 613	4 -	11 11	7 6	13 6	21 9	18 25	23 13	33 24	36 28
172	Malignant melanoma of skin	M F	1,650 2,501	- -	- 1	1 1	2 5	5 18	27 71	67 97	77 163	73 139
173	Other malignant neoplasm of skin	M F	19,087 17,658	1 -	1 5	9 11	14 22	21 35	28 68	97 131	124 186	212 294
174	Malignant neoplasm of female breast	F	31,843	-	-	-	1	6	29	178	510	1,013
175	Malignant neoplasm of male breast	M	191	-	-	-	-	-	-	-	1	3
179	Malignant neoplasm of uterus, part unspecified	F	425	-	1	-	-	1	2	3	3	12
180	Malignant neoplasm of cervix uteri	F	3,400	-	-	-	-	3	34	191	331	388
181	Malignant neoplasm of placenta	F	9	-	-	-	-	-	1	3	3	-
182	Malignant neoplasm of body of uterus	F	3,912	-	-	-	-	1	1	6	18	30
183	Malignant neoplasm of ovary and other uterine adnexa	F	5,388	-	1	2	1	11	35	38	61	110
184	Malignant neoplasm of other and unspecified female genital organs	F	1,148	1	1	-	-	-	2	6	12	26
185	Malignant neoplasm of prostate	M	15,705	-	-	-	-	2	1	-	2	1
186	Malignant neoplasm of testis	M	1,382	1	3	1	2	30	144	273	271	195
187	Malignant neoplasm of penis and other male genital organs	M	359	-	-	-	1	-	-	-	4	8
188	Malignant neoplasm of bladder	M F	8,536 3,476	- -	1 1	- 1	- 1	- -	8 3	6 7	20 12	30 16
189	Malignant neoplasm of kidney and other and unspecified urinary organs	M F	2,793 1,671	7 3	21 17	6 4	1 1	2 4	3 3	7 10	18 16	40 20
190	Malignant neoplasm of eye	M F	248 220	9 5	14 13	4 2	2 1	- -	3 3	6 2	8 7	4 6

40-44	45-49	50-54	55-59	60-64	65-69	70-74	75-79	80-84	85 and over		Site description	ICD (9th Revision) number
2	2	10	16	23	31	52	43	42	37	M	Malignant neoplasm of other and ill-defined sites within the digestive organs and peritoneum	159
2	5	11	13	16	34	32	47	50	101	F		
11	14	13	32	25	23	40	26	21	10	M	Malignant neoplasm of nasal cavities, middle ear and accessory sinuses	160
3	6	9	9	20	32	28	26	26	16	F		
35	85	131	167	255	308	320	193	117	58	M	Malignant neoplasm of larynx	161
10	14	20	37	45	70	65	41	39	18	F		
196	472	935	1,639	2,892	4,798	5,202	4,301	2,952	1,498	M	Malignant neoplasm of trachea, bronchus and lung	162
157	305	440	691	1,481	2,223	2,400	2,071	1,572	913	F		
17	53	89	118	135	166	135	122	55	26	M	Malignant neoplasm of pleura	163
5	10	15	11	20	43	45	37	21	9	F		
7	7	2	14	13	14	15	4	4	2	M	Malignant neoplasm of thymus, heart and mediastinum	164
-	5	3	7	4	10	8	6	9	2	F		
-	-	-	-	-	1	1	-	-	-	M	Other malignant neoplasms within the respiratory system and intrathoracic organs	165
-	1	1	-	-	-	2	-	-	1	F		
17	5	12	14	16	21	20	12	9	6	M	Malignant neoplasm of bone and articular cartilage	170
7	8	7	5	11	8	17	18	12	9	F		
38	46	49	50	55	83	65	53	47	29	M	Malignant neoplasm of connective and other soft tissue	171
44	40	39	36	54	70	60	51	58	39	F		
112	159	161	162	171	192	150	134	103	54	M	Malignant melanoma of skin	172
196	238	181	175	215	223	251	211	163	153	F		
386	708	912	1,331	1,994	2,794	3,318	3,138	2,446	1,553	M	Other malignant neoplasm of skin	173
456	686	785	985	1,356	1,905	2,485	2,937	2,735	2,576	F		
1,908	3,005	3,338	3,589	4,168	3,319	3,215	2,800	2,440	2,324	F	Malignant neoplasm of female breast	174
5	8	10	16	16	26	30	31	25	20	M	Malignant neoplasm of male breast	175
27	41	35	34	28	48	49	39	42	60	F	Malignant neoplasm of uterus, part unspecified	179
325	319	225	231	247	289	295	253	160	109	F	Malignant neoplasm of cervix uteri	180
1	1	-	-	-	-	-	-	-	-	F	Malignant neoplasm of placenta	181
87	177	334	501	549	616	521	449	340	282	F	Malignant neoplasm of body of uterus	182
216	383	455	591	691	737	730	590	429	307	F	Malignant neoplasm of ovary and other uterine adnexa	183
42	40	43	47	64	110	168	190	200	196	F	Malignant neoplasm of other and unspecified female genital organs	184
9	42	149	427	1,105	2,237	3,219	3,613	2,937	1,961	M	Malignant neoplasm of prostate	185
159	111	61	42	29	16	18	10	7	9	M	Malignant neoplasm of testis	186
15	26	27	35	39	39	61	37	39	28	M	Malignant neoplasm of penis and other male genital organs	187
109	188	329	523	986	1,416	1,603	1,493	1,159	665	M	Malignant neoplasm of bladder	188
25	65	104	176	299	462	572	587	615	530	F		
76	133	183	274	369	438	470	364	265	116	M	Malignant neoplasm of kidney and other and unspecified urinary organs	189
43	65	98	116	184	251	250	259	184	143	F		
8	18	12	24	18	33	32	22	27	4	M	Malignant neoplasm of eye	190
14	9	13	16	22	25	28	28	17	9	F		

Series MB1 no. 25 Table 1

Table 1 Series MB1 no. 25

Table 1 Registrations - *continued*

ICD (9th Revision) number	Site description		All ages	Age group								
				Under 1	1-4	5-9	10-14	15-19	20-24	25-29	30-34	35-39
191	Malignant neoplasm of brain	M F	2,029 1,562	13 7	29 33	40 39	44 34	40 23	45 24	55 45	55 57	69 49
192	Malignant neoplasm of other and unspecified parts of nervous system	M F	76 74	- 1	2 3	3 2	1 3	1 2	- 1	3 2	6 4	6 3
193	Malignant neoplasm of thyroid gland	M F	243 713	- -	- 1	1 2	1 -	3 16	7 31	19 43	10 33	10 38
194	Malignant neoplasm of other endocrine glands and related structures	M F	136 123	10 2	21 11	5 4	7 2	4 4	4 1	3 5	2 6	6 6
195	Malignant neoplasm of other and ill-defined sites	M F	134 276	2 3	- 3	1 -	- -	1 -	2 -	4 1	4 3	4 4
196	Secondary and unspecified malignant neoplasm of lymph nodes	M F	235 209	- -	- -	- -	- 1	1 -	- -	3 -	3 2	10 2
197	Secondary malignant neoplasm of respiratory and digestive systems	M F	1,774 1,867	- -	1 -	- -	- -	- -	1 1	2 5	6 9	13 9
198	Secondary malignant neoplasm of other specified sites	M F	840 900	- -	- -	- -	- -	- -	1 1	4 3	6 4	4 7
199	Malignant neoplasm without specification of site	M F	3,212 3,608	1 -	- 2	- -	2 -	2 5	3 4	12 8	22 6	27 13
200,202	Non-Hodgkin's lymphoma	M F	3,702 3,186	4 1	14 11	23 5	22 11	30 19	35 23	74 50	102 39	114 61
200	Lymphosarcoma and reticulosarcoma	M F	270 177	- -	2 1	5 -	3 5	6 4	4 3	8 2	7 1	4 2
201	Hodgkin's disease	M F	729 513	- -	3 -	13 2	24 12	43 47	86 78	88 69	78 52	57 39
202	Other malignant neoplasm of lymphoid and histiocytic tissue	M F	3,432 3,009	4 1	12 10	18 5	19 6	24 15	31 20	66 48	95 38	110 59
203	Multiple myeloma and immunoproliferative neoplasms	M F	1,502 1,358	- -	- -	1 -	- -	- -	- -	4 -	5 -	10 6
204-208	All leukaemias	M F	2,927 2,382	12 8	116 79	55 48	40 21	49 38	60 36	43 31	53 42	54 62
204	Lymphoid leukaemia	M F	1,330 970	8 4	96 70	44 38	24 11	27 15	19 9	13 5	12 11	11 8
205	Myeloid leukaemia	M F	1,382 1,216	3 4	12 6	7 8	16 9	19 21	40 24	27 24	37 30	41 49
206	Monocytic leukaemia	M F	45 36	1 -	- 1	- -	- 1	- -	- -	- -	2 -	- 2
207	Other specified leukaemia	M F	14 10	- -	- -	- -	- -	- -	- -	- -	- -	1 -
208	Leukaemia of unspecified cell type	M F	156 150	- -	8 2	4 2	- -	3 2	1 3	3 2	2 1	1 3
223.3	Benign neoplasm of bladder	M F	49 22	- -	- -	- -	1 -	- -	- -	- -	- 1	1 1
225	Benign neoplasm of brain and other parts of nervous system	M F	448 878	1 -	3 1	1 1	- 3	8 7	9 19	17 15	17 36	29 52
227.3	Benign neoplasm of pituitary gland and craniopharyngeal duct	M F	257 237	- -	- -	- -	1 1	4 3	9 16	20 15	17 29	14 24
227.4	Benign neoplasm of pineal gland	M F	- -	- -	- -	- -	- -	- -	- -	- -	- -	- -
230	Carcinoma in situ of digestive organs	M F	141 109	- -	- -	- -	- -	- -	- -	1 3	1 1	1 3

40-44	45-49	50-54	55-59	60-64	65-69	70-74	75-79	80-84	85 and over		Site description	ICD (9th Revision) number
124	153	173	207	270	273	232	134	50	23	M	Malignant neoplasm of brain	191
76	109	103	126	184	220	173	131	90	39	F		
2	3	7	9	4	8	6	8	5	2	M	Malignant neoplasm of other and unspecified parts of nervous system	192
5	5	5	7	8	2	6	6	5	4	F		
18	17	20	24	20	35	25	18	13	2	M	Malignant neoplasm of thyroid gland	193
46	60	62	57	51	49	64	71	48	41	F		
8	8	4	6	10	17	10	7	3	1	M	Malignant neoplasm of other endocrine glands and related structures	194
6	8	9	8	8	13	11	9	5	5	F		
3	4	9	6	13	20	21	12	16	12	M	Malignant neoplasm of other and ill-defined sites	195
5	4	6	10	13	25	32	47	46	74	F		
10	5	10	19	33	41	41	25	24	10	M	Secondary and unspecified malignant neoplasm of lymph nodes	196
10	17	12	32	17	28	20	27	20	21	F		
21	39	74	115	206	268	335	338	217	138	M	Secondary malignant neoplasm of respiratory and digestive systems	197
26	35	55	91	187	259	289	317	317	267	F		
20	25	51	61	116	172	158	116	75	31	M	Secondary malignant neoplasm of other specified sites	198
10	26	41	49	104	137	165	143	119	91	F		
42	58	97	141	328	493	520	599	493	372	M	Malignant neoplasm without specification of site	199
32	47	82	128	230	360	550	635	690	816	F		
167	234	252	339	402	477	498	445	310	160	M	Non-Hodgkin's lymphoma	200,202
113	132	198	227	332	377	477	462	378	270	F		
12	16	15	20	21	41	36	33	24	13	M	Lymphosarcoma and reticulosarcoma	200
3	5	13	11	22	22	38	21	17	7	F		
63	62	34	28	43	38	26	26	10	7	M	Hodgkin's disease	201
28	23	17	22	21	28	28	25	11	11	F		
155	218	237	319	381	436	462	412	286	147	M	Other malignant neoplasm of lymphoid and histiocytic tissue	202
110	127	185	216	310	355	439	441	361	263	F		
27	40	73	109	200	230	280	234	164	125	M	Multiple myeloma and immunoproliferative neoplasms	203
20	23	44	83	134	185	206	248	217	192	F		
77	87	105	178	247	354	432	424	334	207	M	All leukaemias	204-208
57	59	75	103	155	228	315	321	323	381	F		
26	27	49	70	111	161	197	186	151	98	M	Lymphoid leukaemia	204
12	10	21	45	64	94	124	120	143	166	F		
50	57	49	102	117	161	215	198	142	89	M	Myeloid leukaemia	205
43	47	51	55	82	120	162	166	153	162	F		
-	1	2	2	4	10	2	11	8	2	M	Monocytic leukaemia	206
2	-	2	-	2	-	5	5	6	10	F		
-	-	1	-	3	5	1	3	-	-	M	Other specified leukaemia	207
-	-	-	-	-	-	2	1	3	4	F		
1	2	4	4	12	17	17	26	33	18	M	Leukaemia of unspecified cell type	208
-	2	1	3	7	14	22	29	18	39	F		
2	-	4	6	4	7	11	6	4	3	M	Benign neoplasm of bladder	223.3
1	1	-	3	2	5	2	5	-	1	F		
44	43	34	50	48	50	44	26	16	8	M	Benign neoplasm of brain and other parts of nervous system	225
69	71	65	89	112	100	80	81	51	26	F		
25	22	21	24	32	28	20	14	5	1	M	Benign neoplasm of pituitary gland and craniopharyngeal duct	227.3
30	20	19	16	16	14	13	11	7	3	F		
-	-	-	-	-	-	-	-	-	-	M	Benign neoplasm of pineal gland	227.4
-	-	-	-	-	-	-	-	-	-	F		
1	7	8	17	16	23	20	21	13	12	M	Carcinoma in situ of digestive organs	230
2	3	6	6	11	15	18	21	8	12	F		

Table 1 Series MB1 no. 25

Table 1 Registrations - *continued*

ICD (9th Revision) number	Site description		All ages	Under 1	1-4	5-9	10-14	15-19	20-24	25-29	30-34	35-39
231	Carcinoma in situ of respiratory system	M F	161 40	- -	- -	- -	- -	- 1	- 1	1 1	- 1	2 1
232	Carcinoma in situ of skin	M F	1,140 2,341	- -	- -	- -	- -	2 -	1 4	7 10	8 12	11 19
233	Carcinoma in situ of breast and genitourinary system	M F	589 20,692	- -	- -	- -	- -	- 294	1 2,740	5 4,661	4 3,931	7 2,767
233.0	Carcinoma in situ of breast	M F	10 1,707	- -	- -	- -	- -	- -	- -	- 1	- 33	- 41
233.1	Carcinoma in situ of cervix uteri	F	18,409	-	-	-	-	291	2,728	4,642	3,857	2,676
234	Carcinoma in situ of other and unspecified sites	M F	6 5	- -	- -	- -	- -	- -	- -	- -	- -	- -
235	Neoplasm of uncertain behaviour of digestive and respiratory systems	M F	483 510	1 -	- -	- -	- 3	8 8	10 16	12 18	8 12	12 13
236	Neoplasm of uncertain behaviour of genitourinary organs	M F	309 1,206	- -	- 1	- 1	- 2	3 39	12 197	13 185	11 118	8 119
237	Neoplasm of uncertain behaviour of endocrine glands and nervous system	M F	386 333	2 3	12 4	11 11	8 7	8 12	11 10	18 13	21 18	21 15
238	Neoplasm of uncertain behaviour of other and unspecified sites and tissues	M F	1,000 965	1 4	2 2	4 3	1 2	1 5	13 18	20 16	21 23	25 19
239.4	Neoplasm of unspecified nature of bladder	M F	41 26	- -	- -	- -	- -	- -	- -	- -	- -	1 -
239.6	Neoplasm of unspecified nature of brain	M F	135 147	- -	- 2	1 -	6 1	1 1	3 -	2 3	5 5	3 3
239.7	Neoplasm of unspecified nature of other parts of nervous system and pituitary gland only	M F	16 25	- -	- -	- -	- -	1 -	- -	- -	- -	2 1
630	Hydatidiform mole	F	245	-	-	-	-	43	48	70	50	15

Series MB1 no. 25 Table 1

40-44	45-49	50-54	55-59	60-64	65-69	70-74	75-79	80-84	85 and over		Site description	ICD (9th Revision) number
2	5	15	12	20	29	31	31	8	5	M	Carcinoma in situ of respiratory system	231
1	1	5	1	3	5	7	6	4	2	F		
19	37	51	76	106	191	226	194	139	72	M	Carcinoma in situ of skin	232
39	54	70	135	216	352	372	412	362	284	F		
9	22	28	51	76	109	95	85	65	32	M	Carcinoma in situ of breast and genitourinary system	233
1,882	1,273	890	782	706	326	188	126	76	50	F		
-	1	-	2	1	-	4	1	-	1	M	Carcinoma in situ of breast	233.0
124	171	338	350	346	129	71	59	24	20	F		
1,701	1,064	517	384	309	134	59	22	15	10	F	Carcinoma in situ of cervix uteri	233.1
-	1	-	-	2	2	-	-	1	-	M	Carcinoma in situ of other and unspecified sites	234
-	1	-	-	-	-	1	2	-	1	F		
15	19	18	33	63	77	83	64	43	17	M	Neoplasm of uncertain behaviour of digestive and respiratory systems	235
21	24	21	37	42	54	64	63	63	51	F		
5	9	16	23	31	48	45	42	31	12	M	Neoplasm of uncertain behaviour of genitourinary organs	236
98	78	55	63	51	56	45	43	27	28	F		
24	28	19	26	29	31	37	33	28	19	M	Neoplasm of uncertain behaviour of endocrine glands and nervous system	237
13	15	15	20	27	28	36	30	34	22	F		
24	44	49	78	100	140	157	139	122	59	M	Neoplasm of uncertain behaviour of other and unspecified sites and tissues	238
41	45	38	39	82	97	110	177	136	108	F		
-	1	1	1	3	5	14	6	5	4	M	Neoplasm of unspecified nature of bladder	239.4
-	-	1	3	-	2	5	6	4	5	F		
8	11	6	6	11	13	19	17	15	8	M	Neoplasm of unspecified nature of brain	239.6
4	12	6	5	10	17	22	30	16	10	F		
1	-	1	1	1	1	2	2	3	1	M	Neoplasm of unspecified nature of other parts of nervous system and pituitary gland only	239.7
1	-	3	2	1	3	8	4	1	1	F		
14	5	-	-	-	-	-	-	-	-	F	Hydatidiform mole	630

Table 2 Series MB1 no. 25

Table 2 Estimated resident population:
sex and age as at 30 June 1992

(Figures in thousands)

Area		All ages	Under 1	1-4	5-9	10-14	15-19	20-24	25-29	30-34	35-39
England and Wales	M	**25,098.6**	355.9	**1,415.5**	**1,674.0**	**1,607.6**	**1,598.6**	**1,984.4**	**2,168.8**	**1,967.8**	**1,711.8**
	F	**26,178.2**	338.6	**1,344.6**	**1,584.9**	**1,518.9**	**1,508.0**	**1,894.5**	**2,086.8**	**1,913.8**	**1,695.6**
England	M	**23,687.7**	336.7	1,336.4	1,576.8	1,513.9	1,506.7	1,878.4	2,061.4	1,867.0	1,619.6
	F	**24,690.7**	320.0	1,269.5	1,492.7	1,430.3	1,421.3	1,794.1	1,982.0	1,813.3	1,602.8
Wales	M	**1,411.0**	19.2	79.1	97.2	93.8	91.9	106.0	107.4	100.7	92.2
	F	**1,487.6**	18.6	75.0	92.2	88.6	86.6	100.4	104.8	100.4	92.8
Regional health authorities											
Northern	M	**1,510.9**	21.1	83.0	103.8	100.4	97.4	114.4	121.5	116.1	104.5
	F	**1,588.0**	19.8	78.9	98.3	94.8	92.6	111.0	119.4	115.4	103.4
Yorkshire	M	**1,810.5**	25.8	103.4	123.6	118.1	119.0	147.7	153.9	139.4	122.2
	F	**1,887.5**	24.7	98.4	117.4	111.4	112.6	140.4	144.4	135.1	119.7
Trent	M	**2,340.5**	32.3	128.7	153.5	147.4	150.6	190.5	196.5	179.5	156.5
	F	**2,404.6**	30.5	122.1	145.1	139.2	141.1	176.9	186.1	172.8	154.8
East Anglian	M	**1,030.3**	13.3	55.2	65.6	65.7	67.0	82.3	82.0	76.0	68.7
	F	**1,059.1**	12.9	53.0	62.3	62.1	62.8	74.6	77.1	72.5	68.7
North West Thames	M	**1,767.3**	26.2	99.6	112.3	107.8	105.2	145.4	182.9	157.5	128.5
	F	**1,826.9**	25.0	94.4	106.7	102.4	99.5	146.7	177.9	151.0	125.3
North East Thames	M	**1,858.7**	28.8	110.2	124.9	116.4	115.4	152.4	182.8	161.1	128.7
	F	**1,939.8**	27.8	105.1	118.4	110.6	109.0	149.3	178.7	155.1	128.7
South East Thames	M	**1,794.6**	26.4	103.0	116.4	111.2	107.2	139.9	167.1	147.7	122.5
	F	**1,918.2**	24.9	98.4	111.1	105.2	103.3	138.4	165.0	145.6	122.7
South West Thames	M	**1,477.0**	20.6	79.7	90.8	89.3	86.8	111.8	139.6	123.1	106.5
	F	**1,565.7**	19.6	76.1	85.3	83.7	82.8	109.8	133.8	119.9	104.8
Wessex	M	**1,466.1**	19.0	78.9	94.0	90.9	92.6	114.5	119.8	111.1	100.4
	F	**1,527.3**	18.2	74.6	88.7	86.3	86.3	103.8	112.0	106.3	97.6
Oxford	M	**1,283.3**	18.8	74.6	87.8	87.1	85.9	103.3	112.9	104.6	94.1
	F	**1,298.1**	17.8	70.5	82.8	81.1	80.2	97.9	107.4	100.5	91.6
South Western	M	**1,614.8**	20.8	85.1	102.5	101.3	102.4	122.2	125.5	116.7	105.9
	F	**1,701.1**	19.4	80.7	96.6	95.3	96.2	111.5	119.7	113.7	106.4
West Midlands	M	**2,603.0**	37.4	149.6	179.2	170.4	171.7	208.2	215.8	195.2	172.9
	F	**2,674.6**	35.8	141.5	169.3	160.9	160.2	196.0	205.7	188.6	170.5
Mersey	M	**1,168.9**	16.7	68.3	83.1	78.5	76.9	89.4	94.3	89.5	78.1
	F	**1,243.6**	15.7	64.5	78.7	74.7	73.3	86.9	94.3	90.2	80.1
North Western	M	**1,961.8**	29.4	117.2	139.4	129.5	128.5	156.6	167.0	149.4	130.1
	F	**2,056.3**	27.9	111.2	132.1	122.8	121.5	150.7	160.5	146.7	128.4

Series MB1 no. 25 Table 2

England and Wales, England, Wales, regional health authorities

40-44	45-49	50-54	55-59	60-64	65-69	70-74	75-79	80-84	85 and over		Area
1,760.5	**1,700.0**	**1,360.9**	**1,281.8**	**1,228.1**	**1,129.3**	**919.9**	**638.1**	**389.8**	**205.9**	M	**England and Wales**
1,756.1	**1,694.9**	**1,362.0**	**1,295.2**	**1,305.5**	**1,293.2**	**1,195.2**	**987.3**	**761.6**	**641.6**	F	
1,661.3	1,605.2	1,281.1	1,206.6	1,154.5	1,057.7	862.3	600.0	367.7	194.3	M	England
1,657.2	1,599.7	1,282.4	1,218.4	1,226.8	1,211.7	1,119.9	927.7	716.5	604.4	F	
99.2	94.8	79.9	75.2	73.6	71.5	57.6	38.1	22.1	11.5	M	Wales
98.9	95.2	79.6	76.8	78.7	81.5	75.3	59.6	45.1	37.2	F	
											Regional health authorities
106.6	99.8	83.1	81.2	79.6	73.8	57.5	35.9	20.8	10.5	M	Northern
106.0	98.0	84.8	83.3	86.3	84.6	75.8	57.6	43.2	35.0	F	
127.1	120.2	96.8	92.7	87.1	82.1	65.5	44.6	27.1	14.1	M	Yorkshire
125.9	118.9	97.2	95.4	95.1	93.9	85.2	70.9	54.6	46.4	F	
164.6	161.4	128.2	120.3	116.4	111.2	89.7	59.8	35.2	18.2	M	Trent
161.7	158.2	126.5	121.4	123.3	122.1	111.6	89.3	67.6	54.4	F	
73.9	72.4	55.9	52.2	51.3	48.5	41.9	29.7	18.6	10.2	M	East Anglian
73.3	71.8	54.7	52.0	54.3	54.6	50.8	42.1	32.2	27.1	F	
122.5	115.6	93.8	89.4	80.5	68.3	53.9	39.8	25.1	13.1	M	North West Thames
123.1	117.1	94.1	88.2	82.1	76.2	69.3	60.8	47.4	39.8	F	
124.8	117.5	94.2	90.4	86.6	77.5	61.9	44.2	27.0	14.0	M	North East Thames
127.5	120.9	95.9	90.5	89.6	87.6	80.0	68.3	52.9	44.1	F	
122.5	118.7	93.5	88.0	85.6	78.4	68.4	49.7	31.3	17.2	M	South East Thames
123.8	121.0	95.4	90.8	93.5	94.3	90.3	77.8	62.1	54.6	F	
107.4	104.1	79.4	74.1	69.7	62.2	51.5	39.4	26.2	14.8	M	South West Thames
108.3	105.3	81.1	75.9	75.0	74.1	70.1	62.8	51.0	46.4	F	
103.7	102.7	78.9	73.9	72.9	68.7	59.0	42.7	27.0	15.2	M	Wessex
103.9	101.7	79.3	75.4	78.7	80.7	76.2	64.1	50.1	43.5	F	
96.0	91.5	69.2	62.4	56.1	48.0	38.5	26.9	16.9	9.0	M	Oxford
95.0	90.0	67.1	59.9	56.8	53.6	48.8	39.6	31.0	26.7	F	
113.6	114.3	89.3	83.9	84.1	80.1	68.5	49.4	31.4	17.6	M	South Western
115.0	114.3	89.9	86.2	90.7	93.5	88.6	73.9	58.4	51.2	F	
180.4	178.8	146.7	135.4	131.1	118.7	94.3	63.2	36.0	18.1	M	West Midlands
178.0	175.4	143.1	134.8	135.4	132.5	122.5	96.5	70.7	57.1	F	
81.6	77.5	64.9	63.1	60.4	52.7	41.3	27.9	16.6	8.1	M	Mersey
82.3	77.9	66.9	64.4	65.4	62.1	56.5	46.3	34.4	28.8	F	
136.5	130.7	107.4	99.7	93.2	87.5	70.2	46.9	28.4	14.1	M	North Western
133.6	129.1	106.6	100.3	100.7	102.0	94.4	77.6	60.9	49.3	F	

Table 3 Series MB1 no. 25

Table 3 Rates per 100,000 population of newly diagnosed cases of cancer: site, sex and age, 1992

ICD (9th Revision) number	Site description		All ages	Under 1	1-4	5-9	10-14	15-19	20-24	25-29	30-34	35-39
	All registrations	M F	532.2 602.4	20.2 11.2	18.6 15.8	11.7 9.6	13.5 10.6	20.9 45.8	30.2 189.7	46.3 288.7	60.1 310.3	83.7 333.5
140-208	All malignant neoplasms	M F	511.7 496.2	18.8 9.2	17.4 15.1	10.7 8.6	12.4 9.3	18.6 18.4	26.8 27.7	41.0 48.6	54.4 88.9	75.6 153.5
140-208 x 173	All malignant neoplasms excluding 173	M F	435.6 428.8	18.5 9.2	17.3 14.7	10.2 7.9	11.6 7.9	17.3 16.1	25.3 24.1	36.5 42.3	48.1 79.2	63.3 136.2
140-149	Lip, mouth and pharynx	M F	9.2 5.0	- -	- 0.1	0.1 0.2	0.2 0.2	0.3 0.6	0.4 0.4	1.4 0.7	1.1 0.7	2.7 1.4
140	Malignant neoplasm of lip	M F	0.8 0.2	- -	- 0.1	- 0.1	- -	- -	- -	0.0 0.0	0.1 0.1	0.1 0.1
141	Malignant neoplasm of tongue	M F	1.9 1.1	- -	- -	- -	0.1 -	0.1 -	0.1 -	0.4 0.2	0.3 0.2	0.4 0.2
142	Malignant neoplasm of major salivary glands	M F	0.9 0.8	- -	- -	- 0.1	- 0.2	- 0.2	0.1 0.1	0.3 0.2	0.2 0.2	0.5 0.6
143	Malignant neoplasm of gum	M F	0.3 0.3	- -	- -	- -	- -	- -	- -	0.1 -	- -	0.1 -	0.1 -
144	Malignant neoplasm of floor of mouth	M F	0.9 0.4	- -	- -	- -	- -	- -	- -	- -	- -	0.1 -
145	Malignant neoplasm of other and unspecified parts of mouth	M F	1.2 0.7	- -	- -	- 0.1	0.1 -	0.1 0.3	0.1 0.1	0.1 -	0.4 0.2	0.6 0.1
146	Malignant neoplasm of oropharynx	M F	1.2 0.4	- -	- -	- -	- -	- -	- 0.1	0.0 0.0	- -	0.3 -
147	Malignant neoplasm of nasopharynx	M F	0.5 0.2	- -	- -	0.1 -	0.1 -	0.1 0.1	0.1 0.2	0.5 0.1	0.2 0.1	0.6 0.2
148	Malignant neoplasm of hypopharynx	M F	1.0 0.6	- -	- -	- -	- -	- -	- -	- -	- -	0.1 0.1
149	Malignant neoplasm of other and ill-defined sites within the lip, oral cavity and pharynx	M F	0.4 0.2	- -	- -	- -	- -	- -	- -	0.0 -	- -	0.1 0.1
150	Malignant neoplasm of oesophagus	M F	13.4 9.1	- -	- -	- -	0.1 -	- -	- -	0.1 0.1	0.4 0.1	0.9 0.5
151	Malignant neoplasm of stomach	M F	25.3 14.8	0.3 -	0.1 -	- -	- -	- -	0.1 0.2	0.3 0.3	0.7 0.7	1.9 1.1
152	Malignant neoplasm of small intestine, including duodenum	M F	1.0 0.9	- -	- -	- -	- -	- -	- -	0.0 0.1	0.1 0.2	0.3 0.2
153,154	Colorectal neoplasms	M F	59.5 56.3	- -	- -	- 0.1	- -	0.2 0.3	0.6 0.4	0.8 0.7	2.6 1.8	5.1 4.6
153	Malignant neoplasm of colon	M F	34.2 38.1	- -	- -	- 0.1	- -	0.1 0.1	0.4 0.3	0.3 0.5	1.4 1.1	2.9 2.9
154	Malignant neoplasm of rectum, rectosigmoid junction and anus	M F	25.3 18.2	- -	- -	- -	- -	0.1 0.1	0.2 0.1	0.5 0.1	1.2 0.7	2.2 1.7
155	Malignant neoplasm of liver and intrahepatic bile ducts	M F	3.7 2.2	- -	0.1 0.1	0.1 -	0.2 0.1	0.1 0.2	0.3 0.2	0.1 0.2	0.2 0.2	0.4 0.3
156	Malignant neoplasm of gallbladder and extrahepatic bile ducts	M F	2.3 2.8	- -	- -	- -	- -	- -	- -	0.0 -	0.1 0.1	0.1 0.2
157	Malignant neoplasm of pancreas	M F	12.1 12.3	- -	- -	- -	- -	0.1 -	- 0.1	0.2 -	0.5 0.2	1.2 0.9
158	Malignant neoplasm of retroperitoneum and peritoneum	M F	0.5 0.5	- -	- -	- -	0.1 -	0.1 -	0.1 0.1	- 0.1	0.2 0.1	- 0.2

England and Wales
Registered by July 1998

40-44	45-49	50-54	55-59	60-64	65-69	70-74	75-79	80-84	85 and over		Site description	ICD (9th Revision) number
140.2	237.1	430.1	701.1	1,170.4	1,866.2	2,585.1	3,417.5	4,161.5	4,705.2	M	All registrations	
373.7	481.6	657.2	832.6	1,108.5	1,295.6	1,567.9	1,904.1	2,206.3	2,367.9	F		
130.0	222.5	410.2	669.5	1,126.3	1,799.5	2,497.7	3,311.0	4,033.7	4,582.3	M	All malignant neoplasms	140-208
247.5	387.0	569.5	739.9	1,010.6	1,212.6	1,486.7	1,801.1	2,102.7	2,273.7	F		
108.1	180.8	343.1	565.7	963.9	1,552.1	2,137.0	2,819.2	3,406.2	3,827.9	M	All malignant neoplasms	140-208
221.5	346.5	511.9	663.8	906.7	1,065.3	1,278.7	1,503.6	1,743.6	1,872.2	F	excluding 173	x 173
4.9	10.3	15.6	20.6	24.2	33.4	33.9	38.4	37.2	42.7	M	Lip, mouth and pharynx	140-149
2.9	3.5	6.4	7.9	11.3	13.5	16.7	15.9	15.1	24.2	F		
0.3	0.4	0.9	1.3	1.5	2.8	5.2	4.4	3.6	5.3	M	Malignant neoplasm of lip	140
0.1	0.1	0.2	0.2	0.5	0.4	0.6	1.0	1.3	1.4	F		
1.1	2.6	3.4	4.4	4.6	6.1	5.5	10.7	7.7	5.3	M	Malignant neoplasm of tongue	141
1.0	0.6	1.4	2.4	2.1	3.5	4.3	3.1	2.8	5.1	F		
0.3	0.6	1.2	1.4	2.6	3.8	3.7	3.8	5.6	6.8	M	Malignant neoplasm of major	142
0.4	0.4	1.1	1.2	1.5	1.9	2.3	2.7	2.2	4.2	F	salivary glands	
-	0.3	0.4	0.4	0.9	1.2	0.9	0.6	1.5	4.4	M	Malignant neoplasm of gum	143
-	0.1	-	0.3	0.8	0.8	1.3	1.1	0.8	1.7	F		
0.3	1.4	2.1	3.0	3.6	3.2	3.0	2.4	2.3	2.4	M	Malignant neoplasm of floor of	144
0.1	0.5	1.0	0.6	1.1	1.3	1.0	1.3	0.9	0.8	F	mouth	
0.9	1.4	1.9	2.5	2.8	3.9	3.3	4.9	4.4	10.2	M	Malignant neoplasm of other and	145
0.2	0.5	0.7	0.9	1.7	2.0	2.2	1.7	2.1	5.3	F	unspecified parts of mouth	
0.7	1.7	3.0	2.6	3.3	5.3	4.8	3.4	3.8	2.9	M	Malignant neoplasm of oropharynx	146
0.6	0.4	1.0	0.6	1.1	0.9	1.3	1.4	0.7	0.8	F		
0.9	0.9	0.7	2.1	0.7	1.1	0.9	1.3	0.5	1.0	M	Malignant neoplasm of nasopharynx	147
0.4	0.4	0.1	0.2	0.5	0.4	0.6	0.2	0.8	1.1	F		
0.3	0.7	1.5	1.9	3.2	4.6	4.5	5.2	5.6	3.9	M	Malignant neoplasm of hypopharynx	148
-	0.3	0.8	0.8	1.3	2.0	2.6	2.5	2.5	1.6	F		
0.2	0.4	0.5	1.0	1.2	1.4	2.2	1.9	2.1	0.5	M	Malignant neoplasm of other and	149
0.2	0.1	0.1	0.5	0.6	0.3	0.6	0.7	1.1	2.2	F	ill-defined sites within the lip, oral cavity and pharynx	
3.2	6.6	12.6	22.2	35.8	51.5	66.7	80.9	90.0	98.6	M	Malignant neoplasm of oesophagus	150
0.7	2.6	5.4	8.6	12.8	20.9	31.8	47.5	56.1	64.5	F		
4.3	8.0	16.9	31.5	57.4	95.8	126.5	172.4	223.2	254.1	M	Malignant neoplasm of stomach	151
2.3	3.4	4.9	10.5	18.7	31.6	48.3	74.0	100.6	127.3	F		
0.7	0.7	1.5	1.6	2.4	3.6	4.2	3.9	6.2	8.3	M	Malignant neoplasm of small	152
0.2	0.2	0.6	0.8	1.8	2.6	3.3	4.4	4.6	3.4	F	intestine, including duodenum	
11.5	25.4	53.9	87.3	144.2	217.5	290.8	388.8	464.1	525.1	M	Colorectal neoplasms	153,154
10.4	20.2	37.5	60.3	99.0	142.1	190.4	250.6	324.3	378.9	F		
6.5	11.8	27.4	46.3	79.0	123.1	165.8	234.9	288.4	334.7	M	Malignant neoplasm of colon	153
6.2	13.0	23.8	38.7	66.0	95.9	128.3	167.0	228.6	265.9	F		
5.0	13.6	26.5	41.0	65.2	94.4	125.0	153.9	175.7	190.4	M	Malignant neoplasm of rectum,	154
4.2	7.2	13.7	21.6	33.0	46.2	62.2	83.6	95.7	113.0	F	rectosigmoid junction and anus	
0.8	1.1	3.7	5.8	9.0	14.0	18.8	21.6	27.2	24.3	M	Malignant neoplasm of liver and	155
0.6	0.9	1.0	2.9	4.7	5.3	6.5	8.9	10.4	15.0	F	intrahepatic bile ducts	
0.3	0.4	1.9	3.0	5.8	8.7	11.6	14.7	21.5	21.4	M	Malignant neoplasm of gallbladder	156
0.5	0.6	2.0	2.9	5.7	6.5	8.6	13.7	17.2	19.6	F	and extrahepatic bile ducts	
2.7	4.1	9.7	18.2	29.3	39.8	58.8	84.5	91.1	126.3	M	Malignant neoplasm of pancreas	157
1.8	3.1	7.0	11.5	18.9	30.6	44.7	59.0	78.9	78.4	F		
0.3	0.5	0.9	1.4	1.1	1.3	1.7	1.3	2.6	0.5	M	Malignant neoplasm of	158
0.2	0.3	0.7	0.5	1.4	1.7	2.0	2.1	1.2	2.3	F	retroperitoneum and peritoneum	

Table 3 Rates per 100,000 population - *continued*

ICD (9th Revision) number	Site description		All ages	Under 1	1-4	5-9	10-14	15-19	20-24	25-29	30-34	35-39
159	Malignant neoplasm of other and ill-defined sites within the digestive organs and peritoneum	M F	1.0 1.2	- -	- -	- -	- -	- -	- -	- 0.0	0.1 -	0.1 0.2
160	Malignant neoplasm of nasal cavities, middle ear and accessory sinuses	M F	0.9 0.7	- -	0.1 0.1	- -	- -	- 0.1	0.2 0.1	0.1 0.1	0.1 0.1	0.3 0.2
161	Malignant neoplasm of larynx	M F	6.7 1.4	- -	- -	- -	- -	- 0.1	- -	- 0.0	0.1 0.1	0.6 0.1
162	Malignant neoplasm of trachea, bronchus and lung	M F	99.5 47.1	0.6 -	- 0.1	- -	0.1 -	0.2 0.1	0.1 0.2	0.5 0.3	1.4 0.7	3.3 2.7
163	Malignant neoplasm of pleura	M F	3.7 0.8	- -	- -	- -	- -	- -	- -	0.0 -	- -	0.2 -
164	Malignant neoplasm of thymus, heart and mediastinum	M F	0.4 0.3	- -	0.1 0.3	0.1 -	- -	0.1 0.1	0.2 0.2	0.2 0.1	0.2 -	0.6 0.2
165	Other malignant neoplasms within the respiratory system and intrathoracic organs	M F	0.0 0.0	- -	- -	- -	- -	- -	- -	- -	- -	- -
170	Malignant neoplasm of bone and articular cartilage	M F	0.9 0.6	- -	0.3 -	0.3 0.2	0.9 1.0	1.8 1.1	1.0 0.7	0.5 0.4	0.5 0.2	0.5 0.3
171	Malignant neoplasm of connective and other soft tissue	M F	2.7 2.3	1.1 -	0.8 0.8	0.4 0.4	0.8 0.4	1.3 0.6	0.9 1.3	1.1 0.6	1.7 1.3	2.1 1.7
172	Malignant melanoma of skin	M F	6.6 9.6	- -	- 0.1	0.1 0.1	0.1 0.3	0.3 1.2	1.4 3.7	3.1 4.6	3.9 8.5	4.3 8.2
173	Other malignant neoplasm of skin	M F	76.0 67.5	0.3 -	0.1 0.4	0.5 0.7	0.9 1.4	1.3 2.3	1.4 3.6	4.5 6.3	6.3 9.7	12.4 17.3
174	Malignant neoplasm of female breast	F	121.6	-	-	-	0.1	0.4	1.5	8.5	26.6	59.7
175	Malignant neoplasm of male breast	M	0.8	-	-	-	-	-	-	-	0.1	0.2
179	Malignant neoplasm of uterus, part unspecified	F	1.6	-	0.1	-	-	0.1	0.1	0.1	0.2	0.7
180	Malignant neoplasm of cervix uteri	F	13.0	-	-	-	-	0.2	1.8	9.2	17.3	22.9
181	Malignant neoplasm of placenta	F	0.0	-	-	-	-	-	0.1	0.1	0.2	-
182	Malignant neoplasm of body of uterus	F	14.9	-	-	-	-	0.1	0.1	0.3	0.9	1.8
183	Malignant neoplasm of ovary and other uterine adnexa	F	20.6	-	0.1	0.1	0.1	0.7	1.8	1.8	3.2	6.5
184	Malignant neoplasm of other and unspecified female genital organs	F	4.4	0.3	0.1	-	-	-	0.1	0.3	0.6	1.5
185	Malignant neoplasm of prostate	M	62.6	-	-	-	-	0.1	0.1	-	0.1	0.1
186	Malignant neoplasm of testis	M	5.5	0.3	0.2	0.1	0.1	1.9	7.3	12.6	13.8	11.4
187	Malignant neoplasm of penis and other male genital organs	M	1.4	-	-	-	0.1	-	-	-	0.2	0.5
188	Malignant neoplasm of bladder	M F	34.0 13.3	- -	0.1 0.1	- 0.1	- 0.1	- -	0.4 0.2	0.3 0.3	1.0 0.6	1.8 0.9
189	Malignant neoplasm of kidney and other and unspecified urinary organs	M F	11.1 6.4	2.0 0.9	1.5 1.3	0.4 0.3	0.1 0.1	0.1 0.3	0.2 0.2	0.3 0.5	0.9 0.8	2.3 1.2
190	Malignant neoplasm of eye	M F	1.0 0.8	2.5 1.5	1.0 1.0	0.2 0.1	0.1 0.1	- -	0.2 0.2	0.3 0.1	0.4 0.4	0.2 0.4

40-44	45-49	50-54	55-59	60-64	65-69	70-74	75-79	80-84	85 and over		Site description	ICD (9th Revision) number
0.1	0.1	0.7	1.2	1.9	2.7	5.7	6.7	10.8	18.0	M	Malignant neoplasm of other and ill-defined sites within the digestive organs and peritoneum	159
0.1	0.3	0.8	1.0	1.2	2.6	2.7	4.8	6.6	15.7	F		
0.6	0.8	1.0	2.5	2.0	2.0	4.3	4.1	5.4	4.9	M	Malignant neoplasm of nasal cavities, middle ear and accessory sinuses	160
0.2	0.4	0.7	0.7	1.5	2.5	2.3	2.6	3.4	2.5	F		
2.0	5.0	9.6	13.0	20.8	27.3	34.8	30.2	30.0	28.2	M	Malignant neoplasm of larynx	161
0.6	0.8	1.5	2.9	3.4	5.4	5.4	4.2	5.1	2.8	F		
11.1	27.8	68.7	127.9	235.5	424.9	565.5	674.0	757.3	727.7	M	Malignant neoplasm of trachea, bronchus and lung	162
8.9	18.0	32.3	53.3	113.4	171.9	200.8	209.8	206.4	142.3	F		
1.0	3.1	6.5	9.2	11.0	14.7	14.7	19.1	14.1	12.6	M	Malignant neoplasm of pleura	163
0.3	0.6	1.1	0.8	1.5	3.3	3.8	3.7	2.8	1.4	F		
0.4	0.4	0.1	1.1	1.1	1.2	1.6	0.6	1.0	1.0	M	Malignant neoplasm of thymus, heart and mediastinum	164
-	0.3	0.2	0.5	0.3	0.8	0.7	0.6	1.2	0.3	F		
-	-	-	-	-	0.1	0.1	-	-	-	M	Other malignant neoplasms within the respiratory system and intrathoracic organs	165
-	0.1	0.1	-	-	-	0.2	-	-	0.2	F		
1.0	0.3	0.9	1.1	1.3	1.9	2.2	1.9	2.3	2.9	M	Malignant neoplasm of bone and articular cartilage	170
0.4	0.5	0.5	0.4	0.8	0.6	1.4	1.8	1.6	1.4	F		
2.2	2.7	3.6	3.9	4.5	7.3	7.1	8.3	12.1	14.1	M	Malignant neoplasm of connective and other soft tissue	171
2.5	2.4	2.9	2.8	4.1	5.4	5.0	5.2	7.6	6.1	F		
6.4	9.4	11.8	12.6	13.9	17.0	16.3	21.0	26.4	26.2	M	Malignant melanoma of skin	172
11.2	14.0	13.3	13.5	16.5	17.2	21.0	21.4	21.4	23.8	F		
21.9	41.6	67.0	103.8	162.4	247.4	360.7	491.8	627.5	754.4	M	Other malignant neoplasm of skin	173
26.0	40.5	57.6	76.0	103.9	147.3	207.9	297.5	359.1	401.5	F		
108.7	177.3	245.1	277.1	319.3	256.7	269.0	283.6	320.4	362.2	F	Malignant neoplasm of female breast	174
0.3	0.5	0.7	1.2	1.3	2.3	3.3	4.9	6.4	9.7	M	Malignant neoplasm of male breast	175
1.5	2.4	2.6	2.6	2.1	3.7	4.1	4.0	5.5	9.4	F	Malignant neoplasm of uterus, part unspecified	179
18.5	18.8	16.5	17.8	18.9	22.3	24.7	25.6	21.0	17.0	F	Malignant neoplasm of cervix uteri	180
0.1	0.1	-	-	-	-	-	-	-	-	F	Malignant neoplasm of placenta	181
5.0	10.4	24.5	38.7	42.1	47.6	43.6	45.5	44.6	44.0	F	Malignant neoplasm of body of uterus	182
12.3	22.6	33.4	45.6	52.9	57.0	61.1	59.8	56.3	47.9	F	Malignant neoplasm of ovary and other uterine adnexa	183
2.4	2.4	3.2	3.6	4.9	8.5	14.1	19.2	26.3	30.5	F	Malignant neoplasm of other and unspecified female genital organs	184
0.5	2.5	10.9	33.3	90.0	198.1	349.9	566.2	753.5	952.6	M	Malignant neoplasm of prostate	185
9.0	6.5	4.5	3.3	2.4	1.4	2.0	1.6	1.8	4.4	M	Malignant neoplasm of testis	186
0.9	1.5	2.0	2.7	3.2	3.5	6.6	5.8	10.0	13.6	M	Malignant neoplasm of penis and other male genital organs	187
6.2	11.1	24.2	40.8	80.3	125.4	174.3	234.0	297.3	323.0	M	Malignant neoplasm of bladder	188
1.4	3.8	7.6	13.6	22.9	35.7	47.9	59.5	80.8	82.6	F		
4.3	7.8	13.4	21.4	30.0	38.8	51.1	57.0	68.0	56.3	M	Malignant neoplasm of kidney and other and unspecified urinary organs	189
2.4	3.8	7.2	9.0	14.1	19.4	20.9	26.2	24.2	22.3	F		
0.5	1.1	0.9	1.9	1.5	2.9	3.5	3.4	6.9	1.9	M	Malignant neoplasm of eye	190
0.8	0.5	1.0	1.2	1.7	1.9	2.3	2.8	2.2	1.4	F		

Table 3 Series MB1 no. 25

Table 3 Rates per 100,000 population - *continued*

ICD (9th Revision) number	Site description		All ages	Age group								
				Under 1	1-4	5-9	10-14	15-19	20-24	25-29	30-34	35-39
191	Malignant neoplasm of brain	M	**8.1**	3.7	2.0	2.4	2.7	2.5	2.3	2.5	2.8	4.0
		F	**6.0**	2.1	2.5	2.5	2.2	1.5	1.3	2.2	3.0	2.9
192	Malignant neoplasm of other and unspecified parts of nervous system	M	**0.3**	-	0.1	0.2	0.1	0.1	-	0.1	0.3	0.4
		F	**0.3**	0.3	0.2	0.1	0.2	0.1	0.1	0.1	0.2	0.2
193	Malignant neoplasm of thyroid gland	M	**1.0**	-	-	0.1	0.1	0.2	0.4	0.9	0.5	0.6
		F	**2.7**	-	0.1	0.1	-	1.1	1.6	2.1	1.7	2.2
194	Malignant neoplasm of other endocrine glands and related structures	M	**0.5**	2.8	1.5	0.3	0.4	0.3	0.2	0.1	0.1	0.4
		F	**0.5**	0.6	0.8	0.3	0.1	0.3	0.1	0.2	0.3	0.4
195	Malignant neoplasm of other and ill-defined sites	M	**0.5**	0.6	-	0.1	-	0.1	0.1	0.2	0.2	0.2
		F	**1.1**	0.9	0.2	-	-	-	0.1	-	0.2	0.2
196	Secondary and unspecified malignant neoplasm of lymph nodes	M	**0.9**	-	-	-	-	0.1	-	0.1	0.2	0.6
		F	**0.8**	-	-	-	0.1	-	-	-	0.1	0.1
197	Secondary malignant neoplasm of respiratory and digestive systems	M	**7.1**	-	0.1	-	-	-	0.1	0.1	0.3	0.8
		F	**7.1**	-	-	-	-	-	0.1	0.2	0.5	0.5
198	Secondary malignant neoplasm of other specified sites	M	**3.3**	-	-	-	-	-	0.1	0.2	0.3	0.2
		F	**3.4**	-	-	-	-	-	0.1	0.1	0.2	0.4
199	Malignant neoplasm without specification of site	M	**12.8**	0.3	-	-	0.1	0.1	0.2	0.6	1.1	1.6
		F	**13.8**	-	0.1	-	-	0.3	0.2	0.4	0.3	0.8
200,202	Non-Hodgkin's lymphoma	M	**14.7**	1.1	1.0	1.4	1.4	1.9	1.8	3.4	5.2	6.7
		F	**12.2**	0.3	0.8	0.3	0.7	1.3	1.2	2.4	2.0	3.6
200	Lymphosarcoma and reticulosarcoma	M	**1.1**	-	0.1	0.3	0.2	0.4	0.2	0.4	0.4	0.2
		F	**0.7**	-	0.1	-	0.3	0.3	0.2	0.1	0.1	0.1
201	Hodgkin's disease	M	**2.9**	-	0.2	0.8	1.5	2.7	4.3	4.1	4.0	3.3
		F	**2.0**	-	-	0.1	0.8	3.1	4.1	3.3	2.7	2.3
202	Other malignant neoplasm of lymphoid and histiocytic tissue	M	**13.7**	1.1	0.8	1.1	1.2	1.5	1.6	3.0	4.8	6.4
		F	**11.5**	0.3	0.7	0.3	0.4	1.0	1.1	2.3	2.0	3.5
203	Multiple myeloma and immunoproliferative neoplasms	M	**6.0**	-	-	0.1	-	-	-	0.2	0.3	0.6
		F	**5.2**	-	-	-	-	-	-	-	-	0.4
204-208	All leukaemias	M	**11.7**	3.4	8.2	3.3	2.5	3.1	3.0	2.0	2.7	3.2
		F	**9.1**	2.4	5.9	3.0	1.4	2.5	1.9	1.5	2.2	3.7
204	Lymphoid leukaemia	M	**5.3**	2.2	6.8	2.6	1.5	1.7	1.0	0.6	0.6	0.6
		F	**3.7**	1.2	5.2	2.4	0.7	1.0	0.5	0.2	0.6	0.5
205	Myeloid leukaemia	M	**5.5**	0.8	0.8	0.4	1.0	1.2	2.0	1.2	1.9	2.4
		F	**4.6**	1.2	0.4	0.5	0.6	1.4	1.3	1.2	1.6	2.9
206	Monocytic leukaemia	M	**0.2**	0.3	-	-	-	-	-	-	0.1	-
		F	**0.1**	-	0.1	-	0.1	-	-	-	-	0.1
207	Other specified leukaemia	M	**0.1**	-	-	-	-	-	-	-	-	0.1
		F	**0.0**	-	-	-	-	-	-	-	-	-
208	Leukaemia of unspecified cell type	M	**0.6**	-	0.6	0.2	-	0.2	0.1	0.1	0.1	0.1
		F	**0.6**	-	0.1	0.1	-	0.1	0.2	0.1	0.1	0.2
223.3	Benign neoplasm of bladder	M	**0.2**	-	-	-	0.1	-	-	-	-	0.1
		F	**0.1**	-	-	-	-	-	-	-	0.1	0.1
225	Benign neoplasm of brain and other parts of nervous system	M	**1.8**	0.3	0.2	0.1	-	0.5	0.5	0.8	0.9	1.7
		F	**3.4**	-	0.1	0.1	0.2	0.5	1.0	0.7	1.9	3.1
227.3	Benign neoplasm of pituitary gland and craniopharyngeal duct	M	**1.0**	-	-	-	0.1	0.3	0.5	0.9	0.9	0.8
		F	**0.9**	-	-	-	0.1	0.2	0.8	0.7	1.5	1.4
227.4	Benign neoplasm of pineal gland	M	**-**	-	-	-	-	-	-	-	-	-
		F	**-**	-	-	-	-	-	-	-	-	-
230	Carcinoma in situ of digestive organs	M	**0.6**	-	-	-	-	-	-	0.0	0.1	0.1
		F	**0.4**	-	-	-	-	-	-	0.1	0.1	0.2

40-44	45-49	50-54	55-59	60-64	65-69	70-74	75-79	80-84	85 and over		Site description	ICD (9th Revision) number
7.0	9.0	12.7	16.1	22.0	24.2	25.2	21.0	12.8	11.2	M	Malignant neoplasm of brain	191
4.3	6.4	7.6	9.7	14.1	17.0	14.5	13.3	11.8	6.1	F		
0.1	0.2	0.5	0.7	0.3	0.7	0.7	1.3	1.3	1.0	M	Malignant neoplasm of other and unspecified parts of nervous system	192
0.3	0.3	0.4	0.5	0.6	0.2	0.5	0.6	0.7	0.6	F		
1.0	1.0	1.5	1.9	1.6	3.1	2.7	2.8	3.3	1.0	M	Malignant neoplasm of thyroid gland	193
2.6	3.5	4.6	4.4	3.9	3.8	5.4	7.2	6.3	6.4	F		
0.5	0.5	0.3	0.5	0.8	1.5	1.1	1.1	0.8	0.5	M	Malignant neoplasm of other endocrine glands and related structures	194
0.3	0.5	0.7	0.6	0.6	1.0	0.9	0.9	0.7	0.8	F		
0.2	0.2	0.7	0.5	1.1	1.8	2.3	1.9	4.1	5.8	M	Malignant neoplasm of other and ill-defined sites	195
0.3	0.2	0.4	0.8	1.0	1.9	2.7	4.8	6.0	11.5	F		
0.6	0.3	0.7	1.5	2.7	3.6	4.5	3.9	6.2	4.9	M	Secondary and unspecified malignant neoplasm of lymph nodes	196
0.6	1.0	0.9	2.5	1.3	2.2	1.7	2.7	2.6	3.3	F		
1.2	2.3	5.4	9.0	16.8	23.7	36.4	53.0	55.7	67.0	M	Secondary malignant neoplasm of respiratory and digestive systems	197
1.5	2.1	4.0	7.0	14.3	20.0	24.2	32.1	41.6	41.6	F		
1.1	1.5	3.7	4.8	9.4	15.2	17.2	18.2	19.2	15.1	M	Secondary malignant neoplasm of other specified sites	198
0.6	1.5	3.0	3.8	8.0	10.6	13.8	14.5	15.6	14.2	F		
2.4	3.4	7.1	11.0	26.7	43.7	56.5	93.9	126.5	180.7	M	Malignant neoplasm without specification of site	199
1.8	2.8	6.0	9.9	17.6	27.8	46.0	64.3	90.6	127.2	F		
9.5	13.8	18.5	26.4	32.7	42.2	54.1	69.7	79.5	77.7	M	Non-Hodgkin's lymphoma	200,202
6.4	7.8	14.5	17.5	25.4	29.2	39.9	46.8	49.6	42.1	F		
0.7	0.9	1.1	1.6	1.7	3.6	3.9	5.2	6.2	6.3	M	Lymphosarcoma and reticulosarcoma	200
0.2	0.3	1.0	0.8	1.7	1.7	3.2	2.1	2.2	1.1	F		
3.6	3.6	2.5	2.2	3.5	3.4	2.8	4.1	2.6	3.4	M	Hodgkin's disease	201
1.6	1.4	1.2	1.7	1.6	2.2	2.3	2.5	1.4	1.7	F		
8.8	12.8	17.4	24.9	31.0	38.6	50.2	64.6	73.4	71.4	M	Other malignant neoplasm of lymphoid and histiocytic tissue	202
6.3	7.5	13.6	16.7	23.7	27.5	36.7	44.7	47.4	41.0	F		
1.5	2.4	5.4	8.5	16.3	20.4	30.4	36.7	42.1	60.7	M	Multiple myeloma and immunoproliferative neoplasms	203
1.1	1.4	3.2	6.4	10.3	14.3	17.2	25.1	28.5	29.9	F		
4.4	5.1	7.7	13.9	20.1	31.3	47.0	66.4	85.7	100.6	M	All leukaemias	204-208
3.2	3.5	5.5	8.0	11.9	17.6	26.4	32.5	42.4	59.4	F		
1.5	1.6	3.6	5.5	9.0	14.3	21.4	29.1	38.7	47.6	M	Lymphoid leukaemia	204
0.7	0.6	1.5	3.5	4.9	7.3	10.4	12.2	18.8	25.9	F		
2.8	3.4	3.6	8.0	9.5	14.3	23.4	31.0	36.4	43.2	M	Myeloid leukaemia	205
2.4	2.8	3.7	4.2	6.3	9.3	13.6	16.8	20.1	25.2	F		
-	0.1	0.1	0.2	0.3	0.9	0.2	1.7	2.1	1.0	M	Monocytic leukaemia	206
0.1	-	0.1	-	0.2	-	0.4	0.5	0.8	1.6	F		
-	-	0.1	-	0.2	0.4	0.1	0.5	-	-	M	Other specified leukaemia	207
-	-	-	-	-	-	0.2	0.1	0.4	0.6	F		
0.1	0.1	0.3	0.3	1.0	1.5	1.8	4.1	8.5	8.7	M	Leukaemia of unspecified cell type	208
-	0.1	0.1	0.2	0.5	1.1	1.8	2.9	2.4	6.1	F		
0.1	-	0.3	0.5	0.3	0.6	1.2	0.9	1.0	1.5	M	Benign neoplasm of bladder	223.3
0.1	0.1	-	0.2	0.2	0.4	0.2	0.5	-	0.2	F		
2.5	2.5	2.5	3.9	3.9	4.4	4.8	4.1	4.1	3.9	M	Benign neoplasm of brain and other parts of nervous system	225
3.9	4.2	4.8	6.9	8.6	7.7	6.7	8.2	6.7	4.1	F		
1.4	1.3	1.5	1.9	2.6	2.5	2.2	2.2	1.3	0.5	M	Benign neoplasm of pituitary gland and craniopharyngeal duct	227.3
1.7	1.2	1.4	1.2	1.2	1.1	1.1	1.1	0.9	0.5	F		
-	-	-	-	-	-	-	-	-	-	M	Benign neoplasm of pineal gland	227.4
-	-	-	-	-	-	-	-	-	-	F		
0.1	0.4	0.6	1.3	1.3	2.0	2.2	3.3	3.3	5.8	M	Carcinoma in situ of digestive organs	230
0.1	0.2	0.4	0.5	0.8	1.2	1.5	2.1	1.1	1.9	F		

Table 3 Series MB1 no. 25

Table 3 Rates per 100,000 population - *continued*

ICD (9th Revision) number	Site description		All ages	Under 1	1-4	5-9	10-14	15-19	20-24	25-29	30-34	35-39
231	Carcinoma in situ of respiratory system	M F	0.6 0.2	- -	- -	- -	- -	- 0.1	- 0.1	0.0 0.0	- 0.1	0.1 0.1
232	Carcinoma in situ of skin	M F	4.5 8.9	- -	- -	- -	- -	0.1 -	0.1 0.2	0.3 0.5	0.4 0.6	0.6 1.1
233	Carcinoma in situ of breast and genitourinary system	M F	2.3 79.0	- -	- -	- -	- -	- 19.5	0.1 144.6	0.2 223.4	0.2 205.4	0.4 163.2
233.0	Carcinoma in situ of breast	M F	0.0 6.5	- -	- -	- -	- -	- -	- -	- 0.0	- 1.7	- 2.4
233.1	Carcinoma in situ of cervix uteri	F	70.3	-	-	-	-	19.3	144.0	222.4	201.5	157.8
234	Carcinoma in situ of other and unspecified sites	M F	0.0 0.0	- -	- -	- -	- -	- -	- -	- -	- -	- -
235	Neoplasm of uncertain behaviour of digestive and respiratory systems	M F	1.9 1.9	0.3 -	- -	- -	- 0.2	0.5 0.5	0.5 0.8	0.6 0.9	0.4 0.6	0.7 0.8
236	Neoplasm of uncertain behaviour of genitourinary organs	M F	1.2 4.6	- -	- 0.1	- 0.1	- 0.1	0.2 2.6	0.6 10.4	0.6 8.9	0.6 6.2	0.5 7.0
237	Neoplasm of uncertain behaviour of endocrine glands and nervous system	M F	1.5 1.3	0.6 0.9	0.8 0.3	0.7 0.7	0.5 0.5	0.5 0.8	0.6 0.5	0.8 0.6	1.1 0.9	1.2 0.9
238	Neoplasm of uncertain behaviour of other and unspecified sites and tissues	M F	4.0 3.7	0.3 1.2	0.1 0.1	0.2 0.2	0.1 0.1	0.1 0.3	0.7 1.0	0.9 0.8	1.1 1.2	1.5 1.1
239.4	Neoplasm of unspecified nature of bladder	M F	0.2 0.1	- -	- -	- -	- -	- -	- -	- -	- -	0.1 -
239.6	Neoplasm of unspecified nature of brain	M F	0.5 0.6	- -	- 0.1	0.1 -	0.4 0.1	0.1 0.1	0.2 -	0.1 0.1	0.3 0.3	0.2 0.2
239.7	Neoplasm of unspecified nature of other parts of nervous system and pituitary gland only	M F	0.1 0.1	- -	- -	- -	- -	0.1 -	- -	- -	- -	0.1 0.1
630	Hydatidiform mole	F	0.9	-	-	-	-	2.9	2.5	3.4	2.6	0.9

40-44	45-49	50-54	55-59	60-64	65-69	70-74	75-79	80-84	85 and over		Site description	ICD (9th Revision) number
0.1	0.3	1.1	0.9	1.6	2.6	3.4	4.9	2.1	2.4	M	Carcinoma in situ of respiratory	231
0.1	0.1	0.4	0.1	0.2	0.4	0.6	0.6	0.5	0.3	F	system	
1.1	2.2	3.7	5.9	8.6	16.9	24.6	30.4	35.7	35.0	M	Carcinoma in situ of skin	232
2.2	3.2	5.1	10.4	16.5	27.2	31.1	41.7	47.5	44.3	F		
0.5	1.3	2.1	4.0	6.2	9.7	10.3	13.3	16.7	15.5	M	Carcinoma in situ of breast and	233
107.2	75.1	65.3	60.4	54.1	25.2	15.7	12.8	10.0	7.8	F	genitourinary system	
-	0.1	-	0.2	0.1	-	0.4	0.2	-	0.5	M	Carcinoma in situ of breast	233.0
7.1	10.1	24.8	27.0	26.5	10.0	5.9	6.0	3.2	3.1	F		
96.9	62.8	38.0	29.6	23.7	10.4	4.9	2.2	2.0	1.6	F	Carcinoma in situ of cervix uteri	233.1
-	0.1	-	-	0.2	0.2	-	-	0.3	-	M	Carcinoma in situ of other and	234
-	0.1	-	-	-	-	0.1	0.2	-	0.2	F	unspecified sites	
0.9	1.1	1.3	2.6	5.1	6.8	9.0	10.0	11.0	8.3	M	Neoplasm of uncertain behaviour	235
1.2	1.4	1.5	2.9	3.2	4.2	5.4	6.4	8.3	7.9	F	of digestive and respiratory systems	
0.3	0.5	1.2	1.8	2.5	4.3	4.9	6.6	8.0	5.8	M	Neoplasm of uncertain behaviour	236
5.6	4.6	4.0	4.9	3.9	4.3	3.8	4.4	3.5	4.4	F	of genitourinary organs	
1.4	1.6	1.4	2.0	2.4	2.7	4.0	5.2	7.2	9.2	M	Neoplasm of uncertain behaviour	237
0.7	0.9	1.1	1.5	2.1	2.2	3.0	3.0	4.5	3.4	F	of endocrine glands and nervous system	
1.4	2.6	3.6	6.1	8.1	12.4	17.1	21.8	31.3	28.7	M	Neoplasm of uncertain behaviour	238
2.3	2.7	2.8	3.0	6.3	7.5	9.2	17.9	17.9	16.8	F	of other and unspecified sites and tissues	
-	0.1	0.1	0.1	0.2	0.4	1.5	0.9	1.3	1.9	M	Neoplasm of unspecified nature of	239.4
-	-	0.1	0.2	-	0.2	0.4	0.6	0.5	0.8	F	bladder	
0.5	0.6	0.4	0.5	0.9	1.2	2.1	2.7	3.8	3.9	M	Neoplasm of unspecified nature of	239.6
0.2	0.7	0.4	0.4	0.8	1.3	1.8	3.0	2.1	1.6	F	brain	
0.1	-	0.1	0.1	0.1	0.1	0.2	0.3	0.8	0.5	M	Neoplasm of unspecified nature of	239.7
0.1	-	0.2	0.2	0.1	0.2	0.7	0.4	0.1	0.2	F	other parts of nervous system and pituitary gland only	
0.8	0.3	-	-	-	-	-	-	-	-	F	Hydatidiform mole	630

Table 4 Series MB1 no. 25

**Table 4 Registrations of newly diagnosed cases of cancer:
site, sex and regional health authority of residence, 1992**

ICD (9th Revision) number	Site description		England and Wales	England	Northern	Yorkshire	Trent	East Anglian	North West Thames	North East Thames
					\multicolumn{6}{c}{Regional health authority of residence}
	All registrations	M	133,584	123,986	7,843	9,735	12,417	5,933	7,809	8,609
		F	157,686	147,261	10,004	11,929	14,030	6,835	9,010	9,782
140-208	All malignant neoplasms	M	128,423	119,473	7,581	9,121	11,978	5,581	7,575	8,338
		F	129,905	121,007	7,579	9,272	11,655	5,312	7,589	8,216
140-208 x 173	All malignant neoplasms excluding 173	M	109,336	101,837	6,631	7,715	10,097	4,338	6,788	7,514
		F	112,247	104,584	6,576	7,944	9,935	4,331	6,908	7,494
140-149	Lip, mouth and pharynx	M	2,317	2,131	190	176	191	97	132	140
		F	1,322	1,227	87	93	119	47	93	92
140	Malignant neoplasm of lip	M	196	175	9	20	14	34	3	7
		F	63	57	1	6	3	5	1	4
141	Malignant neoplasm of tongue	M	475	444	35	36	39	14	21	26
		F	299	285	17	17	33	13	23	25
142	Malignant neoplasm of major salivary glands	M	236	211	16	18	11	9	13	23
		F	213	194	19	23	16	3	16	13
143	Malignant neoplasm of gum	M	69	61	4	4	9	3	6	5
		F	71	66	1	6	7	3	1	3
144	Malignant neoplasm of floor of mouth	M	235	213	32	28	22	5	12	9
		F	99	92	7	7	7	3	8	11
145	Malignant neoplasm of other and unspecified parts of mouth	M	298	283	25	21	19	5	20	16
		F	187	171	8	10	13	7	10	13
146	Malignant neoplasm of oropharynx	M	309	284	29	15	23	8	16	18
		F	107	94	8	11	12	2	6	9
147	Malignant neoplasm of nasopharynx	M	138	128	6	14	16	2	14	15
		F	64	63	7	3	10	3	5	5
148	Malignant neoplasm of hypopharynx	M	258	240	25	11	32	14	20	17
		F	157	148	13	7	13	8	13	5
149	Malignant neoplasm of other and ill-defined sites within the lip, oral cavity and pharynx	M	103	92	9	9	6	3	7	4
		F	62	57	6	3	5	-	10	4
150	Malignant neoplasm of oesophagus	M	3,357	3,164	214	193	348	110	209	197
		F	2,380	2,209	125	143	251	80	119	123
151	Malignant neoplasm of stomach	M	6,344	5,852	415	461	644	245	381	445
		F	3,885	3,595	271	307	368	110	228	250
152	Malignant neoplasm of small intestine, including duodenum	M	249	231	12	21	28	10	11	16
		F	233	208	11	26	14	9	11	15
153,154	Colorectal neoplasms	M	14,930	13,843	876	1,078	1,416	659	827	883
		F	14,734	13,732	817	1,008	1,279	676	789	906
153	Malignant neoplasm of colon	M	8,573	7,981	510	597	756	384	509	519
		F	9,973	9,291	564	651	827	462	530	610
154	Malignant neoplasm of rectum, rectosigmoid junction and anus	M	6,357	5,862	366	481	660	275	318	364
		F	4,761	4,441	253	357	452	214	259	296
155	Malignant neoplasm of liver and intrahepatic bile ducts	M	919	813	41	48	91	14	72	99
		F	570	503	26	38	64	11	37	46
156	Malignant neoplasm of gallbladder and extrahepatic bile ducts	M	578	546	48	37	55	29	42	38
		F	743	694	67	47	64	34	44	33
157	Malignant neoplasm of pancreas	M	3,025	2,826	204	180	283	104	191	238
		F	3,213	2,990	216	209	271	125	221	219
158	Malignant neoplasm of retroperitoneum and peritoneum	M	115	106	10	9	12	3	6	9
		F	141	132	6	11	11	4	9	10

**England and Wales,
England, Wales, and
regional health authorities
Registered by July 1998**

South East Thames	South West Thames	Wessex	Oxford	South Western	West Midlands	Mersey	North Western	Wales		Site description	ICD (9th Revision) number
8,927	7,322	8,264	6,673	9,854	14,005	6,315	10,280	9,598	M	All registrations	
10,859	8,547	10,110	8,459	11,075	15,917	7,894	12,810	10,425	F		
8,691	7,092	7,850	6,498	9,406	13,641	6,172	9,949	8,950	M	All malignant neoplasms	140-208
9,641	7,498	8,159	6,572	9,609	13,021	6,424	10,460	8,898	F		
7,823	6,207	7,296	4,816	7,658	11,346	5,083	8,525	7,499	M	All malignant neoplasms	140-208
8,586	6,673	7,730	5,204	8,067	10,927	5,238	8,971	7,663	F	excluding 173	x 173
137	103	160	82	132	216	139	236	186	M	Lip, mouth and pharynx	140-149
97	72	91	57	81	111	71	116	95	F		
5	4	18	15	20	4	2	20	21	M	Malignant neoplasm of lip	140
6	2	7	6	5	3	-	8	6	F		
28	29	32	24	31	55	30	44	31	M	Malignant neoplasm of tongue	141
19	14	32	13	15	25	9	30	14	F		
21	7	25	7	16	18	13	14	25	M	Malignant neoplasm of major	142
16	13	12	12	14	14	11	12	19	F	salivary glands	
8	4	4	-	-	9	2	3	8	M	Malignant neoplasm of gum	143
7	5	5	2	6	8	3	9	5	F		
8	14	8	5	9	29	11	21	22	M	Malignant neoplasm of floor of	144
5	6	7	3	5	7	10	6	7	F	mouth	
22	12	21	13	22	31	22	34	15	M	Malignant neoplasm of other and	145
20	11	16	8	11	13	14	17	16	F	unspecified parts of mouth	
22	12	20	8	17	21	25	50	25	M	Malignant neoplasm of oropharynx	146
8	3	1	4	5	10	5	10	13	F		
7	5	10	3	6	15	5	10	10	M	Malignant neoplasm of nasopharynx	147
4	4	1	2	2	5	5	7	1	F		
9	7	14	4	8	28	18	33	18	M	Malignant neoplasm of hypopharynx	148
12	11	7	4	11	22	11	11	9	F		
7	9	8	3	3	6	11	7	11	M	Malignant neoplasm of other and	149
-	3	3	3	7	4	3	6	5	F	ill-defined sites within the lip, oral cavity and pharynx	
248	212	231	138	226	376	173	289	193	M	Malignant neoplasm of oesophagus	150
188	124	155	105	185	264	128	219	171	F		
424	293	363	236	409	704	340	492	492	M	Malignant neoplasm of stomach	151
284	199	230	125	271	388	219	345	290	F		
14	11	23	9	13	32	9	22	18	M	Malignant neoplasm of small	152
12	13	23	11	15	25	7	16	25	F	intestine, including duodenum	
1,021	855	1,050	641	1,049	1,680	638	1,170	1,087	M	Colorectal neoplasms	153,154
1,050	934	1,095	679	1,178	1,500	729	1,092	1,002	F		
564	537	649	367	600	973	363	653	592	M	Malignant neoplasm of colon	153
713	673	780	467	761	1,006	484	763	682	F		
457	318	401	274	449	707	275	517	495	M	Malignant neoplasm of rectum,	154
337	261	315	212	417	494	245	329	320	F	rectosigmoid junction and anus	
74	46	67	24	48	68	46	75	106	M	Malignant neoplasm of liver and	155
44	29	43	19	41	25	21	59	67	F	intrahepatic bile ducts	
40	22	32	20	41	88	22	32	32	M	Malignant neoplasm of gallbladder	156
40	37	41	39	54	97	40	57	49	F	and extrahepatic bile ducts	
235	184	194	130	227	279	120	257	199	M	Malignant neoplasm of pancreas	157
262	193	205	162	208	300	142	257	223	F		
3	8	12	4	5	11	5	9	9	M	Malignant neoplasm of	158
9	14	17	5	8	11	2	15	9	F	retroperitoneum and peritoneum	

Table 4 Series MB1 no. 25

Table 4 Registrations in regional health authorities - *continued*

ICD (9th Revision) number	Site description		England and Wales	England	Northern	Yorkshire	Trent	East Anglian	North West Thames	North East Thames
159	Malignant neoplasm of other and ill-defined sites within the digestive organs and peritoneum	M F	**260** **315**	248 302	20 17	9 17	15 28	8 6	20 23	31 40
160	Malignant neoplasm of nasal cavities, middle ear and accessory sinuses	M F	**231** **185**	215 168	18 7	22 18	21 14	11 12	15 10	16 10
161	Malignant neoplasm of larynx	M F	**1,681** **365**	1,552 337	139 29	121 39	135 28	52 12	94 22	105 25
162	Malignant neoplasm of trachea, bronchus and lung	M F	**24,985** **12,327**	23,363 11,571	1,910 1,007	1,814 958	2,360 1,027	871 385	1,539 806	1,822 909
163	Malignant neoplasm of pleura	M F	**921** **216**	879 206	90 13	65 16	48 18	29 8	56 16	89 24
164	Malignant neoplasm of thymus, heart and mediastinum	M F	**107** **69**	98 61	7 3	7 4	3 11	4 3	6 7	8 3
165	Other malignant neoplasms within the respiratory system and intrathoracic organs	M F	**2** **5**	2 4	- -	- -	- -	1 -	- -	- -
170	Malignant neoplasm of bone and articular cartilage	M F	**231** **167**	209 144	10 6	13 8	17 11	5 4	19 18	23 12
171	Malignant neoplasm of connective and other soft tissue	M F	**681** **613**	625 574	40 30	52 31	44 49	31 30	46 24	33 31
172	Malignant melanoma of skin	M F	**1,650** **2,501**	1,536 2,340	59 99	110 200	142 193	89 115	87 148	110 138
173	Other malignant neoplasm of skin	M F	**19,087** **17,658**	17,636 16,423	950 1,003	1,406 1,328	1,881 1,720	1,243 981	787 681	824 722
174	Malignant neoplasm of female breast	F	**31,843**	29,618	1,683	2,100	2,883	1,259	2,101	2,177
175	Malignant neoplasm of male breast	M	**191**	179	10	16	13	9	8	13
179	Malignant neoplasm of uterus, part unspecified	F	**425**	366	21	33	44	6	1	-
180	Malignant neoplasm of cervix uteri	F	**3,400**	3,156	242	292	287	124	184	217
181	Malignant neoplasm of placenta	F	**9**	9	1	-	1	-	1	1
182	Malignant neoplasm of body of uterus	F	**3,912**	3,664	179	261	320	181	252	278
183	Malignant neoplasm of ovary and other uterine adnexa	F	**5,388**	5,033	274	359	500	226	357	375
184	Malignant neoplasm of other and unspecified female genital organs	F	**1,148**	1,064	70	91	124	42	82	55
185	Malignant neoplasm of prostate	M	**15,705**	14,755	677	1,119	1,461	819	1,008	1,041
186	Malignant neoplasm of testis	M	**1,382**	1,304	73	97	116	59	81	83
187	Malignant neoplasm of penis and other male genital organs	M	**359**	332	18	32	26	15	20	21
188	Malignant neoplasm of bladder	M F	**8,536** **3,476**	7,945 3,229	437 189	531 208	804 365	238 89	503 201	571 211
189	Malignant neoplasm of kidney and other and unspecified urinary organs	M F	**2,793** **1,671**	2,584 1,546	149 113	210 133	275 173	108 56	180 104	186 87
190	Malignant neoplasm of eye	M F	**248** **220**	238 209	18 13	19 18	16 19	7 5	27 14	19 13

South East Thames	South West Thames	Wessex	Oxford	South Western	West Midlands	Mersey	North Western	Wales		Site description	ICD (9th Revision) number
40	26	16	10	10	-	5	38	12	M	Malignant neoplasm of other and ill-defined sites within the digestive organs and peritoneum	159
58	28	23	8	10	1	2	41	13	F		
12	14	16	6	14	23	10	17	16	M	Malignant neoplasm of nasal cavities, middle ear and accessory sinuses	160
12	6	12	11	13	21	10	12	17	F		
88	77	99	65	77	196	106	198	129	M	Malignant neoplasm of larynx	161
32	15	19	12	18	34	21	31	28	F		
1,835	1,314	1,359	1,005	1,431	2,655	1,340	2,108	1,622	M	Malignant neoplasm of trachea, bronchus and lung	162
913	697	661	467	746	1,089	761	1,145	756	F		
76	49	87	32	68	68	53	69	42	M	Malignant neoplasm of pleura	163
22	14	16	10	14	11	9	15	10	F		
15	4	11	4	6	14	3	6	9	M	Malignant neoplasm of thymus, heart and mediastinum	164
4	2	7	-	1	5	3	3	8	F		
-	-	-	-	1	-	-	-	-	M	Other malignant neoplasms within the respiratory system and intrathoracic organs	165
-	-	3	-	1	-	-	-	1	F		
11	16	20	9	15	15	12	24	22	M	Malignant neoplasm of bone and articular cartilage	170
12	9	17	7	9	12	3	16	23	F		
101	32	54	35	34	66	26	31	56	M	Malignant neoplasm of connective and other soft tissue	171
121	27	59	20	39	69	15	29	39	F		
82	103	143	108	191	147	68	97	114	M	Malignant melanoma of skin	172
178	163	253	114	287	204	83	165	161	F		
868	885	554	1,682	1,748	2,295	1,089	1,424	1,451	M	Other malignant neoplasm of skin	173
1,055	825	429	1,368	1,542	2,094	1,186	1,489	1,235	F		
2,408	1,903	2,205	1,641	2,283	3,178	1,337	2,460	2,225	F	Malignant neoplasm of female breast	174
24	9	17	10	14	20	6	10	12	M	Malignant neoplasm of male breast	175
87	1	46	6	33	48	4	36	59	F	Malignant neoplasm of uterus, part unspecified	179
241	159	182	141	201	344	185	357	244	F	Malignant neoplasm of cervix uteri	180
-	-	-	-	2	2	-	1	-	F	Malignant neoplasm of placenta	181
271	232	289	193	298	424	197	289	248	F	Malignant neoplasm of body of uterus	182
467	345	367	255	359	535	242	372	355	F	Malignant neoplasm of ovary and other uterine adnexa	183
117	57	79	49	61	115	34	88	84	F	Malignant neoplasm of other and unspecified female genital organs	184
1,128	975	1,203	787	1,344	1,563	637	993	950	M	Malignant neoplasm of prostate	185
88	89	109	90	101	145	72	101	78	M	Malignant neoplasm of testis	186
24	19	24	16	22	43	11	41	27	M	Malignant neoplasm of penis and other male genital organs	187
617	537	655	394	688	865	405	700	591	M	Malignant neoplasm of bladder	188
251	229	280	144	249	342	168	303	247	F		
199	172	179	116	216	281	124	189	209	M	Malignant neoplasm of kidney and other and unspecified urinary organs	189
122	96	132	64	120	148	69	129	125	F		
27	15	24	9	11	21	6	19	10	M	Malignant neoplasm of eye	190
24	17	15	19	12	20	5	15	11	F		

Table 4 Registrations in regional health authorities - *continued*

ICD (9th Revision) number	Site description		England and Wales	England	Regional health authority of residence					
					Northern	Yorkshire	Trent	East Anglian	North West Thames	North East Thames
191	Malignant neoplasm of brain	M F	**2,029** **1,562**	1,865 1,452	117 106	133 106	184 140	80 73	128 82	131 104
192	Malignant neoplasm of other and unspecified parts of nervous system	M F	**76** **74**	68 66	1 3	6 3	6 11	2 3	5 2	2 1
193	Malignant neoplasm of thyroid gland	M F	**243** **713**	226 673	16 41	15 43	17 45	7 29	18 43	16 39
194	Malignant neoplasm of other endocrine glands and related structures	M F	**136** **123**	126 116	6 2	9 8	6 7	5 3	11 6	5 6
195	Malignant neoplasm of other and ill-defined sites	M F	**134** **276**	97 223	6 20	5 26	12 20	3 9	2 2	5 11
196	Secondary and unspecified malignant neoplasm of lymph nodes	M F	**235** **209**	226 197	13 9	37 39	27 28	9 10	18 10	21 12
197	Secondary malignant neoplasm of respiratory and digestive systems	M F	**1,774** **1,867**	1,691 1,788	112 118	153 150	216 201	63 54	128 122	139 145
198	Secondary malignant neoplasm of other specified sites	M F	**840** **900**	803 865	52 63	77 76	103 89	24 23	71 67	82 92
199	Malignant neoplasm without specification of site	M F	**3,212** **3,608**	3,023 3,414	195 229	236 296	238 275	151 166	184 185	249 233
200,202	Non-Hodgkin's lymphoma	M F	**3,702** **3,186**	3,418 2,986	169 162	267 223	263 223	172 137	293 206	282 254
200	Lymphosarcoma and reticulosarcoma	M F	**270** **177**	255 169	15 15	31 25	22 9	17 12	15 8	29 14
201	Hodgkin's disease	M F	**729** **513**	677 475	46 28	56 51	45 45	24 19	61 38	54 31
202	Other malignant neoplasm of lymphoid and histiocytic tissue	M F	**3,432** **3,009**	3,163 2,817	154 147	236 198	241 214	155 125	278 198	253 240
203	Multiple myeloma and immunoproliferative neoplasms	M F	**1,502** **1,358**	1,385 1,247	86 63	107 104	154 110	70 54	91 61	108 89
204-208	All leukaemias	M F	**2,927** **2,382**	2,656 2,191	127 110	174 151	262 205	101 92	198 162	184 177
204	Lymphoid leukaemia	M F	**1,330** **970**	1,206 881	62 43	83 57	125 91	45 36	68 51	80 73
205	Myeloid leukaemia	M F	**1,382** **1,216**	1,251 1,130	61 54	83 86	123 107	53 54	108 93	90 91
206	Monocytic leukaemia	M F	**45** **36**	40 34	1 5	1 -	2 1	2 -	1 1	3 1
207	Other specified leukaemia	M F	**14** **10**	13 9	- 2	1 1	3 -	- -	- -	- -
208	Leukaemia of unspecified cell type	M F	**156** **150**	146 137	3 6	6 7	9 6	1 2	21 17	11 12
223.3	Benign neoplasm of bladder	M F	**49** **22**	35 8	2 1	- -	6 3	- -	- -	- -
225	Benign neoplasm of brain and other parts of nervous system	M F	**448** **878**	424 826	29 46	31 58	36 85	24 32	33 80	34 51
227.3	Benign neoplasm of pituitary gland and craniopharyngeal duct	M F	**257** **237**	237 214	11 12	13 17	39 25	9 9	22 19	14 10
227.4	Benign neoplasm of pineal gland	M F	**-** **-**	- -	- -	- -	- -	- -	- -	- -
230	Carcinoma in situ of digestive organs	M F	**141** **109**	118 93	15 6	9 6	14 10	6 6	7 4	4 2

Series MB1 no. 25 Table 4

South East Thames	South West Thames	Wessex	Oxford	South Western	West Midlands	Mersey	North Western	Wales		Site description	ICD (9th Revision) number
137	136	128	104	163	193	90	141	164	M	Malignant neoplasm of brain	191
107	94	106	102	111	151	64	106	110	F		
7	4	10	7	3	4	9	2	8	M	Malignant neoplasm of other and unspecified parts of nervous system	192
3	2	4	10	8	4	6	6	8	F		
26	9	18	13	21	20	8	22	17	M	Malignant neoplasm of thyroid gland	193
48	40	67	46	60	79	28	65	40	F		
6	10	12	5	11	25	5	10	10	M	Malignant neoplasm of other endocrine glands and related structures	194
19	8	7	4	9	25	4	8	7	F		
19	4	16	4	14	-	4	3	37	M	Malignant neoplasm of other and ill-defined sites	195
29	13	33	11	27	3	14	5	53	F		
23	15	19	5	15	3	11	10	9	M	Secondary and unspecified malignant neoplasm of lymph nodes	196
18	12	8	4	24	4	13	6	12	F		
135	112	102	38	122	15	161	195	83	M	Secondary malignant neoplasm of respiratory and digestive systems	197
151	140	124	54	159	17	164	189	79	F		
72	50	48	13	57	3	67	84	37	M	Secondary malignant neoplasm of other specified sites	198
92	70	49	22	44	6	92	80	35	F		
190	149	133	183	161	655	55	244	189	M	Malignant neoplasm without specification of site	199
233	168	160	229	222	676	79	263	194	F		
270	232	277	188	273	344	136	252	284	M	Non-Hodgkin's lymphoma	200,202
225	244	264	131	268	278	117	254	200	F		
13	20	42	11	14	3	14	9	15	M	Lymphosarcoma and reticulosarcoma	200
14	16	25	4	7	1	5	14	8	F		
45	47	43	32	62	83	19	60	52	M	Hodgkin's disease	201
27	34	48	32	27	42	11	42	38	F		
257	212	235	177	259	341	122	243	269	M	Other malignant neoplasm of lymphoid and histiocytic tissue	202
211	228	239	127	261	277	112	240	192	F		
108	84	111	84	111	142	47	82	117	M	Multiple myeloma and immunoproliferative neoplasms	203
133	86	92	82	119	104	53	97	111	F		
222	170	231	160	252	283	95	197	271	M	All leukaemias	204-208
175	147	203	114	192	215	86	162	191	F		
102	58	107	75	139	123	45	94	124	M	Lymphoid leukaemia	204
76	48	101	40	89	72	44	60	89	F		
104	92	106	73	91	130	45	92	131	M	Myeloid leukaemia	205
87	84	79	65	82	119	36	93	86	F		
6	2	2	8	5	4	2	1	5	M	Monocytic leukaemia	206
-	3	6	4	3	6	4	-	2	F		
-	2	4	-	-	2	-	1	1	M	Other specified leukaemia	207
2	-	2	-	-	2	-	-	1	F		
10	16	12	4	17	24	3	9	10	M	Leukaemia of unspecified cell type	208
10	12	15	5	18	16	2	9	13	F		
-	-	12	1	13	-	1	-	14	M	Benign neoplasm of bladder	223.3
-	-	1	2	1	-	-	-	14	F		
33	36	51	24	37	17	9	30	24	M	Benign neoplasm of brain and other parts of nervous system	225
65	77	85	55	78	22	26	66	52	F		
15	8	28	11	26	9	14	18	20	M	Benign neoplasm of pituitary gland and craniopharyngeal duct	227.3
19	14	23	13	15	4	8	26	23	F		
-	-	-	-	-	-	-	-	-	M	Benign neoplasm of pineal gland	227.4
-	-	-	-	-	-	-	-	-	F		
5	5	14	-	13	4	7	15	23	M	Carcinoma in situ of digestive organs	230
3	4	12	5	15	4	3	13	16	F		

Table 4 Registrations in regional health authorities - *continued*

ICD (9th Revision) number	Site description		England and Wales	England	Regional health authority of residence					
					Northern	Yorkshire	Trent	East Anglian	North West Thames	North East Thames
231	Carcinoma in situ of respiratory system	M F	**161** **40**	149 38	2 -	11 6	22 3	7 3	6 3	9 1
232	Carcinoma in situ of skin	M F	**1,140** **2,341**	1,081 2,190	55 135	134 358	129 251	70 166	49 93	47 75
233	Carcinoma in situ of breast and genitourinary system	M F	**589** **20,692**	535 19,759	73 1,469	274 2,005	28 1,731	20 1,137	13 1,055	12 1,284
233.0	Carcinoma in situ of breast	M F	**10** **1,707**	8 1,614	- 62	1 106	- 163	- 82	- 100	1 82
233.1	Carcinoma in situ of cervix uteri	F	**18,409**	17,630	1,354	1,768	1,525	1,033	945	1,169
234	Carcinoma in situ of other and unspecified sites	M F	**6** **5**	5 4	- -	- -	1 -	- -	- -	- -
235	Neoplasm of uncertain behaviour of digestive and respiratory systems	M F	**483** **510**	429 458	24 17	19 29	34 34	28 27	26 38	27 32
236	Neoplasm of uncertain behaviour of genitourinary organs	M F	**309** **1,206**	246 1,152	1 673	2 44	15 45	146 79	12 31	15 38
237	Neoplasm of uncertain behaviour of endocrine glands and nervous system	M F	**386** **333**	338 292	15 13	13 7	30 32	10 7	22 26	46 26
238	Neoplasm of uncertain behaviour of other and unspecified sites and tissues	M F	**1,000** **965**	769 808	18 35	97 95	65 80	32 36	44 58	63 40
239.4	Neoplasm of unspecified nature of bladder	M F	**41** **26**	12 8	2 1	1 -	2 -	- -	- -	- -
239.6	Neoplasm of unspecified nature of brain	M F	**135** **147**	120 139	14 11	9 12	16 15	- -	- -	- -
239.7	Neoplasm of unspecified nature of other parts of nervous system and pituitary gland only	M F	**16** **25**	15 20	1 -	1 -	2 5	- 2	- -	- -
630	Hydatidiform mole	F	**245**	245	6	20	56	19	14	7

South East Thames	South West Thames	Wessex	Oxford	South Western	West Midlands	Mersey	North Western	Wales		Site description	ICD (9th Revision) number
20	15	3	11	3	3	18	19	12	M	Carcinoma in situ of respiratory system	231
2	4	4	-	5	-	3	4	2	F		
37	48	26	90	127	169	-	100	59	M	Carcinoma in situ of skin	232
100	69	56	156	248	291	1	191	151	F		
13	15	16	16	22	4	12	17	54	M	Carcinoma in situ of breast and genitourinary system	233
858	758	1,386	1,624	826	2,408	1,361	1,857	933	F		
-	1	1	1	2	1	-	-	2	M	Carcinoma in situ of breast	233.0
156	153	151	82	111	158	82	126	93	F		
689	586	1,204	1,532	688	2,210	1,249	1,678	779	F	Carcinoma in situ of cervix uteri	233.1
-	1	2	-	-	-	1	-	1	M	Carcinoma in situ of other and unspecified sites	234
-	1	1	1	1	-	-	-	1	F		
22	28	77	1	47	23	32	41	54	M	Neoplasm of uncertain behaviour of digestive and respiratory systems	235
32	28	95	2	57	16	13	38	52	F		
12	7	5	-	5	11	-	15	63	M	Neoplasm of uncertain behaviour of genitourinary organs	236
29	15	73	1	36	5	16	67	54	F		
41	29	26	2	27	52	11	14	48	M	Neoplasm of uncertain behaviour of endocrine glands and nervous system	237
45	34	23	1	29	36	2	11	41	F		
38	38	132	19	112	41	34	36	231	M	Neoplasm of uncertain behaviour of other and unspecified sites and tissues	238
54	35	123	20	116	54	32	30	157	F		
-	-	1	-	2	-	-	4	29	M	Neoplasm of unspecified nature of bladder	239.4
-	-	2	-	1	-	-	4	18	F		
-	-	18	-	11	30	4	18	15	M	Neoplasm of unspecified nature of brain	239.6
-	-	24	-	10	49	5	13	8	F		
-	-	3	-	3	1	-	4	1	M	Neoplasm of unspecified nature of other parts of nervous system and pituitary gland only	239.7
-	-	8	-	2	-	-	3	5	F		
11	10	35	7	26	7	-	27	-	F	Hydatidiform mole	630

Series MB1 no. 25 Table 4

Table 5 Series MB1 no. 25

Table 5 Rates per 100,000 population of newly diagnosed cases of cancer: site, sex and regional health authority of residence, 1992

ICD (9th Revision) number	Site description		England and Wales	England	Northern	Yorkshire	Trent	East Anglian	North West Thames	North East Thames
					Regional health authority of residence					
	All registrations	M F	532.2 602.4	523.4 596.4	519.1 630.0	537.7 632.0	530.5 583.5	575.8 645.4	441.9 493.2	463.2 504.3
140-208	All malignant neoplasms	M F	511.7 496.2	504.4 490.1	501.8 477.3	503.8 491.2	511.8 484.7	541.7 501.6	428.6 415.4	448.6 423.5
140-208 x 173	All malignant neoplasms excluding 173	M F	435.6 428.8	429.9 423.6	438.9 414.1	426.1 420.9	431.4 413.2	421.0 408.9	384.1 378.1	404.3 386.3
140-149	Lip, mouth and pharynx	M F	9.2 5.0	9.0 5.0	12.6 5.5	9.7 4.9	8.2 4.9	9.4 4.4	7.5 5.1	7.5 4.7
140	Malignant neoplasm of lip	M F	0.8 0.2	0.7 0.2	0.6 0.1	1.1 0.3	0.6 0.1	3.3 0.5	0.2 0.1	0.4 0.2
141	Malignant neoplasm of tongue	M F	1.9 1.1	1.9 1.2	2.3 1.1	2.0 0.9	1.7 1.4	1.4 1.2	1.2 1.3	1.4 1.3
142	Malignant neoplasm of major salivary glands	M F	0.9 0.8	0.9 0.8	1.1 1.2	1.0 1.2	0.5 0.7	0.9 0.3	0.7 0.9	1.2 0.7
143	Malignant neoplasm of gum	M F	0.3 0.3	0.3 0.3	0.3 0.1	0.2 0.3	0.4 0.3	0.3 0.3	0.3 0.1	0.3 0.2
144	Malignant neoplasm of floor of mouth	M F	0.9 0.4	0.9 0.4	2.1 0.4	1.5 0.4	0.9 0.3	0.5 0.3	0.7 0.4	0.5 0.6
145	Malignant neoplasm of other and unspecified parts of mouth	M F	1.2 0.7	1.2 0.7	1.7 0.5	1.2 0.5	0.8 0.5	0.5 0.7	1.1 0.5	0.9 0.7
146	Malignant neoplasm of oropharynx	M F	1.2 0.4	1.2 0.4	1.9 0.5	0.8 0.6	1.0 0.5	0.8 0.2	0.9 0.3	1.0 0.5
147	Malignant neoplasm of nasopharynx	M F	0.5 0.2	0.5 0.3	0.4 0.4	0.8 0.2	0.7 0.4	0.2 0.3	0.8 0.3	0.8 0.3
148	Malignant neoplasm of hypopharynx	M F	1.0 0.6	1.0 0.6	1.7 0.8	0.6 0.4	1.4 0.5	1.4 0.8	1.1 0.7	0.9 0.3
149	Malignant neoplasm of other and ill-defined sites within the lip, oral cavity and pharynx	M F	0.4 0.2	0.4 0.2	0.6 0.4	0.5 0.2	0.3 0.2	0.3 -	0.4 0.5	0.2 0.2
150	Malignant neoplasm of oesophagus	M F	13.4 9.1	13.4 8.9	14.2 7.9	10.7 7.6	14.9 10.4	10.7 7.6	11.8 6.5	10.6 6.3
151	Malignant neoplasm of stomach	M F	25.3 14.8	24.7 14.6	27.5 17.1	25.5 16.3	27.5 15.3	23.8 10.4	21.6 12.5	23.9 12.9
152	Malignant neoplasm of small intestine, including duodenum	M F	1.0 0.9	1.0 0.8	0.8 0.7	1.2 1.4	1.2 0.6	1.0 0.8	0.6 0.6	0.9 0.8
153,154	Colorectal neoplasms	M F	59.5 56.3	58.4 55.6	58.0 51.4	59.5 53.4	60.5 53.2	64.0 63.8	46.8 43.2	47.5 46.7
153	Malignant neoplasm of colon	M F	34.2 38.1	33.7 37.6	33.8 35.5	33.0 34.5	32.3 34.4	37.3 43.6	28.8 29.0	27.9 31.4
154	Malignant neoplasm of rectum, rectosigmoid junction and anus	M F	25.3 18.2	24.7 18.0	24.2 15.9	26.6 18.9	28.2 18.8	26.7 20.2	18.0 14.2	19.6 15.3
155	Malignant neoplasm of liver and intrahepatic bile ducts	M F	3.7 2.2	3.4 2.0	2.7 1.6	2.7 2.0	3.9 2.7	1.4 1.0	4.1 2.0	5.3 2.4
156	Malignant neoplasm of gallbladder and extrahepatic bile ducts	M F	2.3 2.8	2.3 2.8	3.2 4.2	2.0 2.5	2.3 2.7	2.8 3.2	2.4 2.4	2.0 1.7
157	Malignant neoplasm of pancreas	M F	12.1 12.3	11.9 12.1	13.5 13.6	9.9 11.1	12.1 11.3	10.1 11.8	10.8 12.1	12.8 11.3
158	Malignant neoplasm of retroperitoneum and peritoneum	M F	0.5 0.5	0.4 0.5	0.7 0.4	0.5 0.6	0.5 0.5	0.3 0.4	0.3 0.5	0.5 0.5

Series MB1 no. 25 Table 5

**England and Wales,
England, Wales, and
regional health authorities
Registered by July 1998**

South East Thames	South West Thames	Wessex	Oxford	South Western	West Midlands	Mersey	North Western	Wales		Site description	ICD (9th Revision) number
497.4	495.7	563.7	520.0	610.2	538.0	540.3	524.0	680.2	M	All registrations	
566.1	545.9	662.0	651.7	651.0	595.1	634.8	623.0	700.8	F		
484.3	480.2	535.4	506.3	582.5	524.0	528.0	507.1	634.3	M	All malignant neoplasms	140-208
502.6	478.9	534.2	506.3	564.9	486.8	516.6	508.7	598.2	F		
435.9	420.2	497.7	375.3	474.2	435.9	434.9	434.6	531.5	M	All malignant neoplasms	140-208
447.6	426.2	506.1	400.9	474.2	408.6	421.2	436.3	515.1	F	excluding 173	x 173
7.6	7.0	10.9	6.4	8.2	8.3	11.9	12.0	13.2	M	Lip, mouth and pharynx	140-149
5.1	4.6	6.0	4.4	4.8	4.2	5.7	5.6	6.4	F		
0.3	0.3	1.2	1.2	1.2	0.2	0.2	1.0	1.5	M	Malignant neoplasm of lip	140
0.3	0.1	0.5	0.5	0.3	0.1	-	0.4	0.4	F		
1.6	2.0	2.2	1.9	1.9	2.1	2.6	2.2	2.2	M	Malignant neoplasm of tongue	141
1.0	0.9	2.1	1.0	0.9	0.9	0.7	1.5	0.9	F		
1.2	0.5	1.7	0.5	1.0	0.7	1.1	0.7	1.8	M	Malignant neoplasm of major	142
0.8	0.8	0.8	0.9	0.8	0.5	0.9	0.6	1.3	F	salivary glands	
0.4	0.3	0.3	-	-	0.3	0.2	0.2	0.6	M	Malignant neoplasm of gum	143
0.4	0.3	0.3	0.2	0.4	0.3	0.2	0.4	0.3	F		
0.4	0.9	0.5	0.4	0.6	1.1	0.9	1.1	1.6	M	Malignant neoplasm of floor of	144
0.3	0.4	0.5	0.2	0.3	0.3	0.8	0.3	0.5	F	mouth	
1.2	0.8	1.4	1.0	1.4	1.2	1.9	1.7	1.1	M	Malignant neoplasm of other and	145
1.0	0.7	1.0	0.6	0.6	0.5	1.1	0.8	1.1	F	unspecified parts of mouth	
1.2	0.8	1.4	0.6	1.1	0.8	2.1	2.5	1.8	M	Malignant neoplasm of oropharynx	146
0.4	0.2	0.1	0.3	0.3	0.4	0.4	0.5	0.9	F		
0.4	0.3	0.7	0.2	0.4	0.6	0.4	0.5	0.7	M	Malignant neoplasm of nasopharynx	147
0.2	0.3	0.1	0.2	0.1	0.2	0.4	0.3	0.1	F		
0.5	0.5	1.0	0.3	0.5	1.1	1.5	1.7	1.3	M	Malignant neoplasm of hypopharynx	148
0.6	0.7	0.5	0.3	0.6	0.8	0.9	0.5	0.6	F		
0.4	0.6	0.5	0.2	0.2	0.2	0.9	0.4	0.8	M	Malignant neoplasm of other and	149
-	0.2	0.2	0.2	0.4	0.1	0.2	0.3	0.3	F	ill-defined sites within the lip, oral cavity and pharynx	
13.8	14.4	15.8	10.8	14.0	14.4	14.8	14.7	13.7	M	Malignant neoplasm of oesophagus	150
9.8	7.9	10.1	8.1	10.9	9.9	10.3	10.7	11.5	F		
23.6	19.8	24.8	18.4	25.3	27.0	29.1	25.1	34.9	M	Malignant neoplasm of stomach	151
14.8	12.7	15.1	9.6	15.9	14.5	17.6	16.8	19.5	F		
0.8	0.7	1.6	0.7	0.8	1.2	0.8	1.1	1.3	M	Malignant neoplasm of small	152
0.6	0.8	1.5	0.8	0.9	0.9	0.6	0.8	1.7	F	intestine, including duodenum	
56.9	57.9	71.6	49.9	65.0	64.5	54.6	59.6	77.0	M	Colorectal neoplasms	153,154
54.7	59.7	71.7	52.3	69.2	56.1	58.6	53.1	67.4	F		
31.4	36.4	44.3	28.6	37.2	37.4	31.1	33.3	42.0	M	Malignant neoplasm of colon	153
37.2	43.0	51.1	36.0	44.7	37.6	38.9	37.1	45.8	F		
25.5	21.5	27.4	21.4	27.8	27.2	23.5	26.4	35.1	M	Malignant neoplasm of rectum,	154
17.6	16.7	20.6	16.3	24.5	18.5	19.7	16.0	21.5	F	rectosigmoid junction and anus	
4.1	3.1	4.6	1.9	3.0	2.6	3.9	3.8	7.5	M	Malignant neoplasm of liver and	155
2.3	1.9	2.8	1.5	2.4	0.9	1.7	2.9	4.5	F	intrahepatic bile ducts	
2.2	1.5	2.2	1.6	2.5	3.4	1.9	1.6	2.3	M	Malignant neoplasm of gallbladder	156
2.1	2.4	2.7	3.0	3.2	3.6	3.2	2.8	3.3	F	and extrahepatic bile ducts	
13.1	12.5	13.2	10.1	14.1	10.7	10.3	13.1	14.1	M	Malignant neoplasm of pancreas	157
13.7	12.3	13.4	12.5	12.2	11.2	11.4	12.5	15.0	F		
0.2	0.5	0.8	0.3	0.3	0.4	0.4	0.5	0.6	M	Malignant neoplasm of	158
0.5	0.9	1.1	0.4	0.5	0.4	0.2	0.7	0.6	F	retroperitoneum and peritoneum	

Table 5 Series MB1 no. 25

Table 5 Rates per 100,000 population - *continued*

ICD (9th Revision) number	Site description		England and Wales	England	Northern	Yorkshire	Trent	East Anglian	North West Thames	North East Thames
					Regional health authority of residence					
159	Malignant neoplasm of other and ill-defined sites within the digestive organs and peritoneum	M F	**1.0** **1.2**	1.0 1.2	1.3 1.1	0.5 0.9	0.6 1.2	0.8 0.6	1.1 1.3	1.7 2.1
160	Malignant neoplasm of nasal cavities, middle ear and accessory sinuses	M F	**0.9** **0.7**	0.9 0.7	1.2 0.4	1.2 1.0	0.9 0.6	1.1 1.1	0.8 0.5	0.9 0.5
161	Malignant neoplasm of larynx	M F	**6.7** **1.4**	6.6 1.4	9.2 1.8	6.7 2.1	5.8 1.2	5.0 1.1	5.3 1.2	5.6 1.3
162	Malignant neoplasm of trachea, bronchus and lung	M F	**99.5** **47.1**	98.6 46.9	126.4 63.4	100.2 50.8	100.8 42.7	84.5 36.4	87.1 44.1	98.0 46.9
163	Malignant neoplasm of pleura	M F	**3.7** **0.8**	3.7 0.8	6.0 0.8	3.6 0.8	2.1 0.7	2.8 0.8	3.2 0.9	4.8 1.2
164	Malignant neoplasm of thymus, heart and mediastinum	M F	**0.4** **0.3**	0.4 0.2	0.5 0.2	0.4 0.2	0.1 0.5	0.4 0.3	0.3 0.4	0.4 0.2
165	Other malignant neoplasms within the respiratory system and intrathoracic organs	M F	**0.0** **0.0**	0.0 0.0	- -	- -	- -	0.1 -	- -	- -
170	Malignant neoplasm of bone and articular cartilage	M F	**0.9** **0.6**	0.9 0.6	0.7 0.4	0.7 0.4	0.7 0.5	0.5 0.4	1.1 1.0	1.2 0.6
171	Malignant neoplasm of connective and other soft tissue	M F	**2.7** **2.3**	2.6 2.3	2.6 1.9	2.9 1.6	1.9 2.0	3.0 2.8	2.6 1.3	1.8 1.6
172	Malignant melanoma of skin	M F	**6.6** **9.6**	6.5 9.5	3.9 6.2	6.1 10.6	6.1 8.0	8.6 10.9	4.9 8.1	5.9 7.1
173	Other malignant neoplasm of skin	M F	**76.0** **67.5**	74.5 66.5	62.9 63.2	77.7 70.4	80.4 71.5	120.6 92.6	44.5 37.3	44.3 37.2
174	Malignant neoplasm of female breast	F	**121.6**	120.0	106.0	111.3	119.9	118.9	115.0	112.2
175	Malignant neoplasm of male breast	M	**0.8**	0.8	0.7	0.9	0.6	0.9	0.5	0.7
179	Malignant neoplasm of uterus, part unspecified	F	**1.6**	1.5	1.3	1.7	1.8	0.6	0.1	-
180	Malignant neoplasm of cervix uteri	F	**13.0**	12.8	15.2	15.5	11.9	11.7	10.1	11.2
181	Malignant neoplasm of placenta	F	**0.0**	0.0	0.1	-	0.0	-	0.1	0.1
182	Malignant neoplasm of body of uterus	F	**14.9**	14.8	11.3	13.8	13.3	17.1	13.8	14.3
183	Malignant neoplasm of ovary and other uterine adnexa	F	**20.6**	20.4	17.3	19.0	20.8	21.3	19.5	19.3
184	Malignant neoplasm of other and unspecified female genital organs	F	**4.4**	4.3	4.4	4.8	5.2	4.0	4.5	2.8
185	Malignant neoplasm of prostate	M	**62.6**	62.3	44.8	61.8	62.4	79.5	57.0	56.0
186	Malignant neoplasm of testis	M	**5.5**	5.5	4.8	5.4	5.0	5.7	4.6	4.5
187	Malignant neoplasm of penis and other male genital organs	M	**1.4**	1.4	1.2	1.8	1.1	1.5	1.1	1.1
188	Malignant neoplasm of bladder	M F	**34.0** **13.3**	33.5 13.1	28.9 11.9	29.3 11.0	34.4 15.2	23.1 8.4	28.5 11.0	30.7 10.9
189	Malignant neoplasm of kidney and other and unspecified urinary organs	M F	**11.1** **6.4**	10.9 6.3	9.9 7.1	11.6 7.0	11.7 7.2	10.5 5.3	10.2 5.7	10.0 4.5
190	Malignant neoplasm of eye	M F	**1.0** **0.8**	1.0 0.8	1.2 0.8	1.0 1.0	0.7 0.8	0.7 0.5	1.5 0.8	1.0 0.7

South East Thames	South West Thames	Wessex	Oxford	South Western	West Midlands	Mersey	North Western	Wales		Site description	ICD (9th Revision) number
2.2	1.8	1.1	0.8	0.6	-	0.4	1.9	0.9	M	Malignant neoplasm of other and ill-defined sites within the digestive organs and peritoneum	159
3.0	1.8	1.5	0.6	0.6	0.0	0.2	2.0	0.9	F		
0.7	0.9	1.1	0.5	0.9	0.9	0.9	0.9	1.1	M	Malignant neoplasm of nasal cavities, middle ear and accessory sinuses	160
0.6	0.4	0.8	0.8	0.8	0.8	0.8	0.6	1.1	F		
4.9	5.2	6.8	5.1	4.8	7.5	9.1	10.1	9.1	M	Malignant neoplasm of larynx	161
1.7	1.0	1.2	0.9	1.1	1.3	1.7	1.5	1.9	F		
102.3	89.0	92.7	78.3	88.6	102.0	114.6	107.5	115.0	M	Malignant neoplasm of trachea, bronchus and lung	162
47.6	44.5	43.3	36.0	43.9	40.7	61.2	55.7	50.8	F		
4.2	3.3	5.9	2.5	4.2	2.6	4.5	3.5	3.0	M	Malignant neoplasm of pleura	163
1.1	0.9	1.0	0.8	0.8	0.4	0.7	0.7	0.7	F		
0.8	0.3	0.8	0.3	0.4	0.5	0.3	0.3	0.6	M	Malignant neoplasm of thymus, heart and mediastinum	164
0.2	0.1	0.5	-	0.1	0.2	0.2	0.4	0.5	F		
-	-	-	-	0.1	-	-	-	-	M	Other malignant neoplasms within the respiratory system and intrathoracic organs	165
-	-	0.2	-	0.1	-	-	-	0.1	F		
0.6	1.1	1.4	0.7	0.9	0.6	1.0	1.2	1.6	M	Malignant neoplasm of bone and articular cartilage	170
0.6	0.6	1.1	0.5	0.5	0.4	0.2	0.8	1.5	F		
5.6	2.2	3.7	2.7	2.1	2.5	2.2	1.6	4.0	M	Malignant neoplasm of connective and other soft tissue	171
6.3	1.7	3.9	1.5	2.3	2.6	1.2	1.4	2.6	F		
4.6	7.0	9.8	8.4	11.8	5.6	5.8	4.9	8.1	M	Malignant melanoma of skin	172
9.3	10.4	16.6	8.8	16.9	7.6	6.7	8.0	10.8	F		
48.4	59.9	37.8	131.1	108.2	88.2	93.2	72.6	102.8	M	Other malignant neoplasm of skin	173
55.0	52.7	28.1	105.4	90.6	78.3	95.4	72.4	83.0	F		
125.5	121.5	144.4	126.4	134.2	118.8	107.5	119.6	149.6	F	Malignant neoplasm of female breast	174
1.3	0.6	1.2	0.8	0.9	0.8	0.5	0.5	0.9	M	Malignant neoplasm of male breast	175
4.5	0.1	3.0	0.5	1.9	1.8	0.3	1.8	4.0	F	Malignant neoplasm of uterus, part unspecified	179
12.6	10.2	11.9	10.9	11.8	12.9	14.9	17.4	16.4	F	Malignant neoplasm of cervix uteri	180
-	-	-	-	0.1	0.1	-	0.0	-	F	Malignant neoplasm of placenta	181
14.1	14.8	18.9	14.9	17.5	15.9	15.8	14.1	16.7	F	Malignant neoplasm of body of uterus	182
24.3	22.0	24.0	19.6	21.1	20.0	19.5	18.1	23.9	F	Malignant neoplasm of ovary and other uterine adnexa	183
6.1	3.6	5.2	3.8	3.6	4.3	2.7	4.3	5.6	F	Malignant neoplasm of other and unspecified female genital organs	184
62.9	66.0	82.1	61.3	83.2	60.0	54.5	50.6	67.3	M	Malignant neoplasm of prostate	185
4.9	6.0	7.4	7.0	6.3	5.6	6.2	5.1	5.5	M	Malignant neoplasm of testis	186
1.3	1.3	1.6	1.2	1.4	1.7	0.9	2.1	1.9	M	Malignant neoplasm of penis and other male genital organs	187
34.4	36.4	44.7	30.7	42.6	33.2	34.6	35.7	41.9	M	Malignant neoplasm of bladder	188
13.1	14.6	18.3	11.1	14.6	12.8	13.5	14.7	16.6	F		
11.1	11.6	12.2	9.0	13.4	10.8	10.6	9.6	14.8	M	Malignant neoplasm of kidney and other and unspecified urinary organs	189
6.4	6.1	8.6	4.9	7.1	5.5	5.5	6.3	8.4	F		
1.5	1.0	1.6	0.7	0.7	0.8	0.5	1.0	0.7	M	Malignant neoplasm of eye	190
1.3	1.1	1.0	1.5	0.7	0.7	0.4	0.7	0.7	F		

Table 5 Series MB1 no. 25

Table 5 Rates per 100,000 population - *continued*

ICD (9th Revision) number	Site description		England and Wales	England	Regional health authority of residence					
					Northern	Yorkshire	Trent	East Anglian	North West Thames	North East Thames
191	Malignant neoplasm of brain	M F	**8.1** **6.0**	7.9 5.9	7.7 6.7	7.3 5.6	7.9 5.8	7.8 6.9	7.2 4.5	7.0 5.4
192	Malignant neoplasm of other and unspecified parts of nervous system	M F	**0.3** **0.3**	0.3 0.3	0.1 0.2	0.3 0.2	0.3 0.5	0.2 0.3	0.3 0.1	0.1 0.1
193	Malignant neoplasm of thyroid gland	M F	**1.0** **2.7**	1.0 2.7	1.1 2.6	0.8 2.3	0.7 1.9	0.7 2.7	1.0 2.4	0.9 2.0
194	Malignant neoplasm of other endocrine glands and related structures	M F	**0.5** **0.5**	0.5 0.5	0.4 0.1	0.5 0.4	0.3 0.3	0.5 0.3	0.6 0.3	0.3 0.3
195	Malignant neoplasm of other and ill-defined sites	M F	**0.5** **1.1**	0.4 0.9	0.4 1.3	0.3 1.4	0.5 0.8	0.3 0.8	0.1 0.1	0.3 0.6
196	Secondary and unspecified malignant neoplasm of lymph nodes	M F	**0.9** **0.8**	1.0 0.8	0.9 0.6	2.0 2.1	1.2 1.2	0.9 0.9	1.0 0.5	1.1 0.6
197	Secondary malignant neoplasm of respiratory and digestive systems	M F	**7.1** **7.1**	7.1 7.2	7.4 7.4	8.5 7.9	9.2 8.4	6.1 5.1	7.2 6.7	7.5 7.5
198	Secondary malignant neoplasm of other specified sites	M F	**3.3** **3.4**	3.4 3.5	3.4 4.0	4.3 4.0	4.4 3.7	2.3 2.2	4.0 3.7	4.4 4.7
199	Malignant neoplasm without specification of site	M F	**12.8** **13.8**	12.8 13.8	12.9 14.4	13.0 15.7	10.2 11.4	14.7 15.7	10.4 10.1	13.4 12.0
200,202	Non-Hodgkin's lymphoma	M F	**14.7** **12.2**	14.4 12.1	11.2 10.2	14.7 11.8	11.2 9.3	16.7 12.9	16.6 11.3	15.2 13.1
200	Lymphosarcoma and reticulosarcoma	M F	**1.1** **0.7**	1.1 0.7	1.0 0.9	1.7 1.3	0.9 0.4	1.6 1.1	0.8 0.4	1.6 0.7
201	Hodgkin's disease	M F	**2.9** **2.0**	2.9 1.9	3.0 1.8	3.1 2.7	1.9 1.9	2.3 1.8	3.5 2.1	2.9 1.6
202	Other malignant neoplasm of lymphoid and histiocytic tissue	M F	**13.7** **11.5**	13.4 11.4	10.2 9.3	13.0 10.5	10.3 8.9	15.0 11.8	15.7 10.8	13.6 12.4
203	Multiple myeloma and immunoproliferative neoplasms	M F	**6.0** **5.2**	5.8 5.1	5.7 4.0	5.9 5.5	6.6 4.6	6.8 5.1	5.1 3.3	5.8 4.6
204-208	All leukaemias	M F	**11.7** **9.1**	11.2 8.9	8.4 6.9	9.6 8.0	11.2 8.5	9.8 8.7	11.2 8.9	9.9 9.1
204	Lymphoid leukaemia	M F	**5.3** **3.7**	5.1 3.6	4.1 2.7	4.6 3.0	5.3 3.8	4.4 3.4	3.8 2.8	4.3 3.8
205	Myeloid leukaemia	M F	**5.5** **4.6**	5.3 4.6	4.0 3.4	4.6 4.6	5.3 4.4	5.1 5.1	6.1 5.1	4.8 4.7
206	Monocytic leukaemia	M F	**0.2** **0.1**	0.2 0.1	0.1 0.3	0.1 -	0.1 0.0	0.2 -	0.1 0.1	0.2 0.1
207	Other specified leukaemia	M F	**0.1** **0.0**	0.1 0.0	- 0.1	0.1 0.1	0.1 -	- -	- -	- -
208	Leukaemia of unspecified cell type	M F	**0.6** **0.6**	0.6 0.6	0.2 0.4	0.3 0.4	0.4 0.2	0.1 0.2	1.2 0.9	0.6 0.6
223.3	Benign neoplasm of bladder	M F	**0.2** **0.1**	0.1 0.0	0.1 0.1	- -	0.3 0.1	- -	- -	- -
225	Benign neoplasm of brain and other parts of nervous system	M F	**1.8** **3.4**	1.8 3.3	1.9 2.9	1.7 3.1	1.5 3.5	2.3 3.0	1.9 4.4	1.8 2.6
227.3	Benign neoplasm of pituitary gland and craniopharyngeal duct	M F	**1.0** **0.9**	1.0 0.9	0.7 0.8	0.7 0.9	1.7 1.0	0.9 0.8	1.2 1.0	0.8 0.5
227.4	Benign neoplasm of pineal gland	M F	**-** **-**	- -	- -	- -	- -	- -	- -	- -
230	Carcinoma in situ of digestive organs	M F	**0.6** **0.4**	0.5 0.4	1.0 0.4	0.5 0.3	0.6 0.4	0.6 0.6	0.4 0.2	0.2 0.1

South East Thames	South West Thames	Wessex	Oxford	South Western	West Midlands	Mersey	North Western	Wales		Site description	ICD (9th Revision) number
7.6	9.2	8.7	8.1	10.1	7.4	7.7	7.2	11.6	M	Malignant neoplasm of brain	191
5.6	6.0	6.9	7.9	6.5	5.6	5.1	5.2	7.4	F		
0.4	0.3	0.7	0.5	0.2	0.2	0.8	0.1	0.6	M	Malignant neoplasm of other and unspecified parts of nervous system	192
0.2	0.1	0.3	0.8	0.5	0.1	0.5	0.3	0.5	F		
1.4	0.6	1.2	1.0	1.3	0.8	0.7	1.1	1.2	M	Malignant neoplasm of thyroid gland	193
2.5	2.6	4.4	3.5	3.5	3.0	2.3	3.2	2.7	F		
0.3	0.7	0.8	0.4	0.7	1.0	0.4	0.5	0.7	M	Malignant neoplasm of other endocrine glands and related structures	194
1.0	0.5	0.5	0.3	0.5	0.9	0.3	0.4	0.5	F		
1.1	0.3	1.1	0.3	0.9	-	0.3	0.2	2.6	M	Malignant neoplasm of other and ill-defined sites	195
1.5	0.8	2.2	0.8	1.6	0.1	1.1	0.2	3.6	F		
1.3	1.0	1.3	0.4	0.9	0.1	0.9	0.5	0.6	M	Secondary and unspecified malignant neoplasm of lymph nodes	196
0.9	0.8	0.5	0.3	1.4	0.1	1.0	0.3	0.8	F		
7.5	7.6	7.0	3.0	7.6	0.6	13.8	9.9	5.9	M	Secondary malignant neoplasm of respiratory and digestive systems	197
7.9	8.9	8.1	4.2	9.3	0.6	13.2	9.2	5.3	F		
4.0	3.4	3.3	1.0	3.5	0.1	5.7	4.3	2.6	M	Secondary malignant neoplasm of other specified sites	198
4.8	4.5	3.2	1.7	2.6	0.2	7.4	3.9	2.4	F		
10.6	10.1	9.1	14.3	10.0	25.2	4.7	12.4	13.4	M	Malignant neoplasm without specification of site	199
12.1	10.7	10.5	17.6	13.1	25.3	6.4	12.8	13.0	F		
15.0	15.7	18.9	14.6	16.9	13.2	11.6	12.8	20.1	M	Non-Hodgkin's lymphoma	200,202
11.7	15.6	17.3	10.1	15.8	10.4	9.4	12.4	13.4	F		
0.7	1.4	2.9	0.9	0.9	0.1	1.2	0.5	1.1	M	Lymphosarcoma and reticulosarcoma	200
0.7	1.0	1.6	0.3	0.4	0.0	0.4	0.7	0.5	F		
2.5	3.2	2.9	2.5	3.8	3.2	1.6	3.1	3.7	M	Hodgkin's disease	201
1.4	2.2	3.1	2.5	1.6	1.6	0.9	2.0	2.6	F		
14.3	14.4	16.0	13.8	16.0	13.1	10.4	12.4	19.1	M	Other malignant neoplasm of lymphoid and histiocytic tissue	202
11.0	14.6	15.6	9.8	15.3	10.4	9.0	11.7	12.9	F		
6.0	5.7	7.6	6.5	6.9	5.5	4.0	4.2	8.3	M	Multiple myeloma and immunoproliferative neoplasms	203
6.9	5.5	6.0	6.3	7.0	3.9	4.3	4.7	7.5	F		
12.4	11.5	15.8	12.5	15.6	10.9	8.1	10.0	19.2	M	All leukaemias	204-208
9.1	9.4	13.3	8.8	11.3	8.0	6.9	7.9	12.8	F		
5.7	3.9	7.3	5.8	8.6	4.7	3.8	4.8	8.8	M	Lymphoid leukaemia	204
4.0	3.1	6.6	3.1	5.2	2.7	3.5	2.9	6.0	F		
5.8	6.2	7.2	5.7	5.6	5.0	3.8	4.7	9.3	M	Myeloid leukaemia	205
4.5	5.4	5.2	5.0	4.8	4.4	2.9	4.5	5.8	F		
0.3	0.1	0.1	0.6	0.3	0.2	0.2	0.1	0.4	M	Monocytic leukaemia	206
-	0.2	0.4	0.3	0.2	0.2	0.3	-	0.1	F		
-	0.1	0.3	-	-	0.1	-	0.1	0.1	M	Other specified leukaemia	207
0.1	-	0.1	-	-	0.1	-	-	0.1	F		
0.6	1.1	0.8	0.3	1.1	0.9	0.3	0.5	0.7	M	Leukaemia of unspecified cell type	208
0.5	0.8	1.0	0.4	1.1	0.6	0.2	0.4	0.9	F		
-	-	0.8	0.1	0.8	-	0.1	-	1.0	M	Benign neoplasm of bladder	223.3
-	-	0.1	0.2	0.1	-	-	-	0.9	F		
1.8	2.4	3.5	1.9	2.3	0.7	0.8	1.5	1.7	M	Benign neoplasm of brain and other parts of nervous system	225
3.4	4.9	5.6	4.2	4.6	0.8	2.1	3.2	3.5	F		
0.8	0.5	1.9	0.9	1.6	0.3	1.2	0.9	1.4	M	Benign neoplasm of pituitary gland and craniopharyngeal duct	227.3
1.0	0.9	1.5	1.0	0.9	0.1	0.6	1.3	1.5	F		
-	-	-	-	-	-	-	-	-	M	Benign neoplasm of pineal gland	227.4
-	-	-	-	-	-	-	-	-	F		
0.3	0.3	1.0	-	0.8	0.2	0.6	0.8	1.6	M	Carcinoma in situ of digestive organs	230
0.2	0.3	0.8	0.4	0.9	0.1	0.2	0.6	1.1	F		

Table 5 Rates per 100,000 population - *continued*

ICD (9th Revision) number	Site description		England and Wales	England	Northern	Yorkshire	Trent	East Anglian	North West Thames	North East Thames
					\multicolumn{6}{l	}{Regional health authority of residence}
231	Carcinoma in situ of respiratory system	M F	**0.6** **0.2**	0.6 0.2	0.1 -	0.6 0.3	0.9 0.1	0.7 0.3	0.3 0.2	0.5 0.1
232	Carcinoma in situ of skin	M F	**4.5** **8.9**	4.6 8.9	3.6 8.5	7.4 19.0	5.5 10.4	6.8 15.7	2.8 5.1	2.5 3.9
233	Carcinoma in situ of breast and genitourinary system	M F	**2.3** **79.0**	2.3 80.0	4.8 92.5	15.1 106.2	1.2 72.0	1.9 107.4	0.7 57.7	0.6 66.2
233.0	Carcinoma in situ of breast	M F	**0.0** **6.5**	0.0 6.5	- 3.9	0.1 5.6	- 6.8	- 7.7	- 5.5	0.1 4.2
233.1	Carcinoma in situ of cervix uteri	F	**70.3**	71.4	85.3	93.7	63.4	97.5	51.7	60.3
234	Carcinoma in situ of other and unspecified sites	M F	**0.0** **0.0**	0.0 0.0	- -	- -	0.0 -	- -	- -	- -
235	Neoplasm of uncertain behaviour of digestive and respiratory systems	M F	**1.9** **1.9**	1.8 1.9	1.6 1.1	1.0 1.5	1.5 1.4	2.7 2.5	1.5 2.1	1.5 1.6
236	Neoplasm of uncertain behaviour of genitourinary organs	M F	**1.2** **4.6**	1.0 4.7	0.1 42.4	0.1 2.3	0.6 1.9	14.2 7.5	0.7 1.7	0.8 2.0
237	Neoplasm of uncertain behaviour of endocrine glands and nervous system	M F	**1.5** **1.3**	1.4 1.2	1.0 0.8	0.7 0.4	1.3 1.3	1.0 0.7	1.2 1.4	2.5 1.3
238	Neoplasm of uncertain behaviour of other and unspecified sites and tissues	M F	**4.0** **3.7**	3.2 3.3	1.2 2.2	5.4 5.0	2.8 3.3	3.1 3.4	2.5 3.2	3.4 2.1
239.4	Neoplasm of unspecified nature of bladder	M F	**0.2** **0.1**	0.1 0.0	0.1 0.1	0.1 -	0.1 -	- -	- -	- -
239.6	Neoplasm of unspecified nature of brain	M F	**0.5** **0.6**	0.5 0.6	0.9 0.7	0.5 0.6	0.7 0.6	- -	- -	- -
239.7	Neoplasm of unspecified nature of other parts of nervous system and pituitary gland only	M F	**0.1** **0.1**	0.1 0.1	0.1 -	0.1 -	0.1 0.2	- 0.2	- -	- -
630	Hydatidiform mole	F	**0.9**	1.0	0.4	1.1	2.3	1.8	0.8	0.4

Series MB1 no. 25 Table 5

South East Thames	South West Thames	Wessex	Oxford	South Western	West Midlands	Mersey	North Western	Wales		Site description	ICD (9th Revision) number
1.1	1.0	0.2	0.9	0.2	0.1	1.5	1.0	0.9	M	Carcinoma in situ of respiratory	231
0.1	0.3	0.3	-	0.3	-	0.2	0.2	0.1	F	system	
2.1	3.2	1.8	7.0	7.9	6.5	-	5.1	4.2	M	Carcinoma in situ of skin	232
5.2	4.4	3.7	12.0	14.6	10.9	0.1	9.3	10.2	F		
0.7	1.0	1.1	1.2	1.4	0.2	1.0	0.9	3.8	M	Carcinoma in situ of breast and	233
44.7	48.4	90.8	125.1	48.6	90.0	109.4	90.3	62.7	F	genitourinary system	
-	0.1	0.1	0.1	0.1	0.0	-	-	0.1	M	Carcinoma in situ of breast	233.0
8.1	9.8	9.9	6.3	6.5	5.9	6.6	6.1	6.3	F		
35.9	37.4	78.8	118.0	40.4	82.6	100.4	81.6	52.4	F	Carcinoma in situ of cervix uteri	233.1
-	0.1	0.1	-	-	-	0.1	-	0.1	M	Carcinoma in situ of other and	234
-	0.1	0.1	0.1	0.1	-	-	-	0.1	F	unspecified sites	
1.2	1.9	5.3	0.1	2.9	0.9	2.7	2.1	3.8	M	Neoplasm of uncertain behaviour	235
1.7	1.8	6.2	0.2	3.4	0.6	1.0	1.8	3.5	F	of digestive and respiratory systems	
0.7	0.5	0.3	-	0.3	0.4	-	0.8	4.5	M	Neoplasm of uncertain behaviour	236
1.5	1.0	4.8	0.1	2.1	0.2	1.3	3.3	3.6	F	of genitourinary organs	
2.3	2.0	1.8	0.2	1.7	2.0	0.9	0.7	3.4	M	Neoplasm of uncertain behaviour	237
2.3	2.2	1.5	0.1	1.7	1.3	0.2	0.5	2.8	F	of endocrine glands and nervous system	
2.1	2.6	9.0	1.5	6.9	1.6	2.9	1.8	16.4	M	Neoplasm of uncertain behaviour	238
2.8	2.2	8.1	1.5	6.8	2.0	2.6	1.5	10.6	F	of other and unspecified sites and tissues	
-	-	0.1	-	0.1	-	-	0.2	2.1	M	Neoplasm of unspecified nature of	239.4
-	-	0.1	-	0.1	-	-	0.2	1.2	F	bladder	
-	-	1.2	-	0.7	1.2	0.3	0.9	1.1	M	Neoplasm of unspecified nature of	239.6
-	-	1.6	-	0.6	1.8	0.4	0.6	0.5	F	brain	
-	-	0.2	-	0.2	0.0	-	0.2	0.1	M	Neoplasm of unspecified nature of	239.7
-	-	0.5	-	0.1	-	-	0.1	0.3	F	other parts of nervous system and pituitary gland only	
0.6	0.6	2.3	0.5	1.5	0.3	-	1.3	-	F	Hydatidiform mole	630

Table 6 Series MB1 no. 25

Table 6 Standardised registration ratios: site, sex and regional health authority of residence, 1992 (England and Wales = 100)

ICD (9th Revision) number	Site description		England	Northern	Yorkshire	Trent	East Anglian	North West Thames	North East Thames	South East Thames
	All registrations	M	99	98	103	98	100	92	93	90
		F	99	105	105	97	104	88	88	91
140-208	All malignant neoplasms	M	99	98	100	99	98	93	94	91
		F	99	96	99	98	98	92	91	98
140-208 x 173	All malignant neoplasms excluding 173	M	99	101	100	98	89	98	99	96
		F	99	96	98	97	92	97	96	101
140-149	Lip, mouth and pharynx	M	98	134	107	87	96	88	87	82
		F	99	108	98	98	85	111	100	97
140	Malignant neoplasm of lip	M	95	76	144	75	392	24	52	35
		F	96	26	132	52	190	25	91	124
141	Malignant neoplasm of tongue	M	99	121	106	87	68	68	78	82
		F	101	93	79	120	104	121	120	84
142	Malignant neoplasm of major salivary glands	M	95	112	107	49	87	85	140	122
		F	97	147	150	82	34	117	87	99
143	Malignant neoplasm of gum	M	94	96	82	138	99	135	104	158
		F	99	23	117	108	100	23	62	129
144	Malignant neoplasm of floor of mouth	M	96	220	167	99	50	78	55	48
		F	99	114	98	76	72	128	161	68
145	Malignant neoplasm of other and unspecified parts of mouth	M	101	139	99	68	39	102	77	102
		F	97	70	74	76	89	85	100	140
146	Malignant neoplasm of oropharynx	M	98	152	68	78	60	80	84	100
		F	93	121	143	121	45	87	121	101
147	Malignant neoplasm of nasopharynx	M	98	71	142	123	35	147	151	71
		F	105	181	65	170	113	118	110	84
148	Malignant neoplasm of hypopharynx	M	99	158	60	130	124	122	95	48
		F	100	134	62	90	121	133	46	101
149	Malignant neoplasm of other and ill-defined sites within the lip, oral cavity and pharynx	M	95	142	123	61	67	106	56	94
		F	98	161	67	89	-	256	93	-
150	Malignant neoplasm of oesophagus	M	100	105	81	109	74	98	85	100
		F	99	88	83	116	80	81	75	102
151	Malignant neoplasm of stomach	M	98	109	103	107	86	95	102	90
		F	98	117	110	105	67	95	93	93
152	Malignant neoplasm of small intestine, including duodenum	M	99	80	119	119	91	69	93	77
		F	95	78	155	66	92	76	93	67
153,154	Colorectal neoplasms	M	99	98	102	100	99	87	86	92
		F	99	92	95	96	109	86	89	92
153	Malignant neoplasm of colon	M	99	99	98	93	100	94	88	88
		F	99	94	91	91	110	86	89	92
154	Malignant neoplasm of rectum, rectosigmoid junction and anus	M	98	95	107	110	97	79	83	98
		F	99	88	104	104	107	88	90	92
155	Malignant neoplasm of liver and intrahepatic bile ducts	M	94	74	74	105	34	123	156	109
		F	94	76	92	123	46	104	117	100
156	Malignant neoplasm of gallbladder and extrahepatic bile ducts	M	100	138	90	101	112	115	95	93
		F	99	150	88	95	109	96	65	69
157	Malignant neoplasm of pancreas	M	99	113	84	99	77	100	114	105
		F	99	112	90	93	92	111	99	105
158	Malignant neoplasm of retroperitoneum and peritoneum	M	98	142	110	110	61	80	112	36
		F	100	69	108	85	68	101	102	84

**England, Wales,
regional health authorities
Registered by July 1998**

South West Thames	Wessex	Oxford	South Western	West Midlands	Mersey	North Western	Wales		Site description	ICD (9th Revision) number
91	97	112	101	104	104	102	120	M	All registrations	
87	104	120	99	101	106	104	112	F		
92	96	114	100	105	106	103	117	M	All malignant neoplasms	140-208
93	100	116	103	101	105	103	115	F		
95	105	99	95	103	103	104	115	M	All malignant neoplasms	140-208
96	110	106	100	98	99	102	114	F	excluding 173	x 173
75	112	77	80	91	130	134	135	M	Lip, mouth and pharynx	140-149
88	110	98	85	84	113	112	120	F		
34	146	171	140	20	22	135	178	M	Malignant neoplasm of lip	140
51	177	219	110	48	-	162	160	F		
103	109	109	92	112	137	122	110	M	Malignant neoplasm of tongue	141
76	172	99	70	84	63	129	78	F		
50	169	65	95	75	120	78	179	M	Malignant neoplasm of major	142
99	91	127	92	66	109	72	150	F	salivary glands	
96	92	-	-	129	64	58	196	M	Malignant neoplasm of gum	143
114	111	67	115	114	89	162	116	F		
102	56	46	55	118	100	117	157	M	Malignant neoplasm of floor of	144
100	114	69	71	70	211	78	118	F	mouth	
67	114	93	104	102	161	150	85	M	Malignant neoplasm of other and	145
95	136	98	81	70	158	116	143	F	unspecified parts of mouth	
66	106	56	78	65	174	211	135	M	Malignant neoplasm of oropharynx	146
46	15	83	66	92	98	120	204	F		
61	122	44	65	105	78	94	126	M	Malignant neoplasm of nasopharynx	147
101	26	67	45	78	166	141	27	F		
46	87	34	43	106	151	168	116	M	Malignant neoplasm of hypopharynx	148
115	71	60	96	140	147	89	95	F		
149	125	64	41	57	231	89	178	M	Malignant neoplasm of other and	149
76	76	112	154	66	103	124	135	F	ill-defined sites within the lip, oral cavity and pharynx	
106	109	92	92	110	113	114	96	M	Malignant neoplasm of oesophagus	150
82	102	105	105	114	115	117	119	F		
77	89	84	87	110	119	103	130	M	Malignant neoplasm of stomach	151
80	92	77	93	103	121	113	124	F		
74	147	79	72	126	79	117	122	M	Malignant neoplasm of small	152
90	156	111	88	108	64	88	178	F	intestine, including duodenum	
96	110	97	95	111	94	104	122	M	Colorectal neoplasms	153,154
101	117	109	108	104	105	95	113	F		
104	118	97	95	113	94	101	115	M	Malignant neoplasm of colon	153
107	122	111	103	103	103	98	114	F		
84	99	97	97	109	95	108	130	M	Malignant neoplasm of rectum,	154
87	104	105	119	105	109	88	112	F	rectosigmoid junction and anus	
84	115	59	72	73	110	108	192	M	Malignant neoplasm of liver and	155
81	119	78	98	45	78	132	196	F	intrahepatic bile ducts	
63	86	79	96	151	84	74	92	M	Malignant neoplasm of gallbladder	156
79	86	125	98	133	115	98	110	F	and extrahepatic bile ducts	
101	100	97	102	91	88	113	110	M	Malignant neoplasm of pancreas	157
95	100	120	87	95	94	102	115	F		
118	170	74	62	93	94	102	132	M	Malignant neoplasm of	158
162	193	82	79	78	30	136	106	F	retroperitoneum and peritoneum	

Table 6 Series MB1 no. 25

Table 6 Standardised registration ratios - *continued*

ICD (9th Revision) number	Site description		England	Northern	Yorkshire	Trent	East Anglian	North West Thames	North East Thames	South East Thames
				\multicolumn{7}{l	}{Regional health authority of residence}
159	Malignant neoplasm of other and ill-defined sites within the digestive organs and peritoneum	M F	101 102	131 92	49 75	61 99	68 45	122 118	173 184	203 233
160	Malignant neoplasm of nasal cavities, middle ear and accessory sinuses	M F	99 97	129 62	134 135	96 82	109 155	100 86	99 78	71 85
161	Malignant neoplasm of larynx	M F	98 98	135 128	101 148	84 83	71 79	88 96	90 99	72 117
162	Malignant neoplasm of trachea, bronchus and lung	M F	100 100	127 133	102 108	100 91	78 74	98 106	106 107	99 98
163	Malignant neoplasm of pleura	M F	102 102	159 97	99 103	55 90	72 88	95 120	140 162	114 135
164	Malignant neoplasm of thymus, heart and mediastinum	M F	97 94	107 71	91 80	30 173	88 105	83 155	105 61	195 77
165	Other malignant neoplasms within the respiratory system and intrathoracic organs	M F	107 85	- -	- -	- -	1129 -	- -	- -	- -
170	Malignant neoplasm of bone and articular cartilage	M F	96 92	72 59	78 66	78 72	51 58	122 161	138 100	67 97
171	Malignant neoplasm of connective and other soft tissue	M F	97 99	97 80	107 70	69 87	106 118	101 59	68 71	204 264
172	Malignant melanoma of skin	M F	99 99	59 65	94 112	91 84	126 111	79 89	94 77	69 96
173	Other malignant neoplasm of skin	M F	98 99	83 94	104 105	104 107	146 132	65 61	62 59	61 78
174	Malignant neoplasm of female breast	F	99	86	92	98	95	102	98	102
175	Malignant neoplasm of male breast	M	100	88	118	72	105	66	98	169
179	Malignant neoplasm of uterus, part unspecified	F	92	82	108	113	34	4	-	271
180	Malignant neoplasm of cervix uteri	F	99	117	120	92	89	79	88	96
181	Malignant neoplasm of placenta	F	106	189	-	122	-	141	138	-
182	Malignant neoplasm of body of uterus	F	100	74	92	89	111	102	103	93
183	Malignant neoplasm of ovary and other uterine adnexa	F	99	82	93	101	101	104	100	116
184	Malignant neoplasm of other and unspecified female genital organs	F	99	102	110	119	87	114	69	132
185	Malignant neoplasm of prostate	M	100	74	101	99	114	103	96	94
186	Malignant neoplasm of testis	M	100	89	98	91	106	78	78	88
187	Malignant neoplasm of penis and other male genital organs	M	98	83	126	77	95	86	84	91
188	Malignant neoplasm of bladder	M F	99 99	85 90	88 83	100 115	62 61	93 93	97 88	97 93
189	Malignant neoplasm of kidney and other and unspecified urinary organs	M F	98 98	88 110	106 111	104 113	88 80	100 99	96 75	98 97
190	Malignant neoplasm of eye	M F	102 101	120 96	107 114	69 94	66 55	164 98	108 83	149 146

South West Thames	Wessex	Oxford	South Western	West Midlands	Mersey	North Western	Wales		Site description	ICD (9th Revision) number
163	94	88	51	-	43	196	78	M	Malignant neoplasm of other and ill-defined sites within the digestive organs and peritoneum	159
137	113	60	42	3	14	167	69	F		
102	111	56	85	97	94	97	117	M	Malignant neoplasm of nasal cavities, middle ear and accessory sinuses	160
53	103	138	97	114	114	83	153	F		
78	94	86	64	113	137	155	128	M	Malignant neoplasm of larynx	161
68	83	76	69	93	121	109	127	F		
88	85	92	78	105	118	112	108	M	Malignant neoplasm of trachea, bronchus and lung	162
92	85	90	83	89	130	118	101	F		
91	152	76	103	72	124	98	76	M	Malignant neoplasm of pleura	163
107	118	108	89	51	88	89	76	F		
63	170	78	82	127	60	73	144	M	Malignant neoplasm of thymus, heart and mediastinum	164
48	166	-	21	72	92	148	197	F		
-	-	-	688	-	-	-	-	M	Other malignant neoplasms within the respiratory system and intrathoracic organs	165
-	957	-	277	-	-	-	335	F		
118	144	79	97	63	112	134	165	M	Malignant neoplasm of bone and articular cartilage	170
89	169	89	79	71	38	122	237	F		
79	130	107	72	95	83	59	142	M	Malignant neoplasm of connective and other soft tissue	171
72	158	71	92	112	52	61	108	F		
104	142	137	168	87	90	77	119	M	Malignant melanoma of skin	172
106	167	98	166	81	70	85	110	F		
77	45	197	124	119	126	99	128	M	Other malignant neoplasm of skin	173
74	38	180	119	120	143	108	117	F		
98	113	114	102	99	88	99	118	F	Malignant neoplasm of female breast	174
78	139	117	99	104	70	70	106	M	Malignant neoplasm of male breast	175
4	174	31	108	113	20	109	234	F	Malignant neoplasm of uterus, part unspecified	179
76	89	87	87	100	115	136	124	F	Malignant neoplasm of cervix uteri	180
-	-	-	369	222	-	144	-	F	Malignant neoplasm of placenta	181
97	119	113	106	108	105	95	105	F	Malignant neoplasm of body of uterus	182
105	110	106	94	98	94	89	110	F	Malignant neoplasm of ovary and other uterine adnexa	183
78	108	99	72	102	63	98	123	F	Malignant neoplasm of other and unspecified female genital organs	184
101	116	116	112	101	92	85	101	M	Malignant neoplasm of prostate	185
105	136	124	117	103	115	95	105	M	Malignant neoplasm of testis	186
88	106	97	85	118	67	151	127	M	Malignant neoplasm of penis and other male genital organs	187
105	120	105	109	101	105	109	116	M	Malignant neoplasm of bladder	188
105	126	98	97	100	103	111	118	F		
104	102	92	107	98	97	89	125	M	Malignant neoplasm of kidney and other and unspecified urinary organs	189
93	126	88	100	89	87	99	124	F		
102	158	77	64	83	53	100	69	M	Malignant neoplasm of eye	190
127	112	190	78	90	48	87	84	F		

Table 6 Series MB1 no. 25

Table 6 Standardised registration ratios - *continued*

ICD (9th Revision) number	Site description		England	Northern	Yorkshire	Trent	East Anglian	North West Thames	North East Thames	South East Thames
				Regional health authority of residence						
191	Malignant neoplasm of brain	M F	98 99	94 110	92 94	96 97	93 113	95 80	91 94	95 93
192	Malignant neoplasm of other and unspecified parts of nervous system	M F	95 95	22 67	111 56	84 162	62 99	97 40	37 19	127 55
193	Malignant neoplasm of thyroid gland	M F	99 100	108 95	86 84	74 69	68 99	110 90	92 76	148 90
194	Malignant neoplasm of other endocrine glands and related structures	M F	98 100	73 27	92 90	47 62	89 60	120 73	50 68	62 209
195	Malignant neoplasm of other and ill-defined sites	M F	77 86	75 123	53 131	95 81	51 78	23 12	53 57	192 134
196	Secondary and unspecified malignant neoplasm of lymph nodes	M F	102 100	91 70	221 259	121 146	87 115	119 75	128 83	134 115
197	Secondary malignant neoplasm of respiratory and digestive systems	M F	101 102	105 104	122 112	129 118	79 69	114 105	114 113	102 105
198	Secondary malignant neoplasm of other specified sites	M F	102 102	101 114	129 117	129 108	65 61	133 120	141 149	117 134
199	Malignant neoplasm without specification of site	M F	100 101	103 107	104 114	79 85	104 109	91 83	112 94	79 82
200,202	Non-Hodgkin's lymphoma	M F	98 100	76 83	101 97	75 76	107 103	121 102	109 115	100 93
200	Lymphosarcoma and reticulosarcoma	M F	100 102	92 137	161 196	86 55	144 162	85 72	153 114	66 106
201	Hodgkin's disease	M F	98 98	105 90	107 138	66 95	80 92	117 104	100 80	86 71
202	Other malignant neoplasm of lymphoid and histiocytic tissue	M F	98 100	74 80	97 91	74 78	104 99	124 104	105 115	103 93
203	Multiple myeloma and immunoproliferative neoplasms	M F	98 98	95 77	101 106	109 89	105 94	95 73	104 95	97 127
204-208	All leukaemias	M F	96 98	73 77	84 88	95 95	78 93	104 107	89 106	102 96
204	Lymphoid leukaemia	M F	96 97	78 74	88 81	100 103	77 89	79 83	85 107	104 102
205	Myeloid leukaemia	M F	96 99	74 74	84 98	95 97	87 107	120 119	93 107	102 94
206	Monocytic leukaemia	M F	94 100	37 236	31 -	47 31	100 -	35 44	95 40	179 -
207	Other specified leukaemia	M F	99 96	- 349	100 139	224 -	- -	- -	- -	- 245
208	Leukaemia of unspecified cell type	M F	99 97	33 68	54 65	62 44	14 32	209 181	101 115	85 85
223.3	Benign neoplasm of bladder	M F	76 39	67 73	- -	129 148	- -	- -	- -	- -
225	Benign neoplasm of brain and other parts of nervous system	M F	100 100	106 85	97 92	85 105	126 88	109 138	107 82	103 100
227.3	Benign neoplasm of pituitary gland and craniopharyngeal duct	M F	98 96	70 83	71 100	161 115	84 93	125 114	76 57	82 109
227.4	Benign neoplasm of pineal gland	M F	- -	- -	- -	- -	- -	- -	- -	- -
230	Carcinoma in situ of digestive organs	M F	89 91	177 90	90 77	105 100	96 132	77 58	41 26	48 36

South West Thames	Wessex	Oxford	South Western	West Midlands	Mersey	North Western	Wales		Site description	ICD (9th Revision) number
115	105	107	118	92	95	90	138	M	Malignant neoplasm of brain	191
100	112	142	103	95	86	87	119	F		
88	217	192	58	51	257	34	182	M	Malignant neoplasm of other and	192
45	90	286	159	53	170	104	186	F	unspecified parts of nervous system	
62	122	112	127	80	71	118	121	M	Malignant neoplasm of thyroid	193
91	155	138	123	110	83	117	97	F	gland	
127	150	74	123	177	78	94	127	M	Malignant neoplasm of other	194
108	95	69	108	200	68	83	98	F	endocrine glands and related structures	
49	190	65	145	-	66	30	470	M	Malignant neoplasm of other and	195
73	186	94	131	11	109	23	322	F	ill-defined sites	
107	129	47	89	13	102	56	64	M	Secondary and unspecified	196
93	62	43	161	19	131	37	96	F	malignant neoplasm of lymph nodes	
105	90	49	93	8	201	146	78	M	Secondary malignant neoplasm of	197
120	104	69	115	9	187	129	70	F	respiratory and digestive systems	
101	91	35	94	3	174	132	73	M	Secondary malignant neoplasm of	198
126	86	57	67	7	216	114	64	F	other specified sites	
76	64	129	67	204	38	102	99	M	Malignant neoplasm without	199
73	69	152	82	193	47	93	90	F	specification of site	
105	120	109	104	91	80	90	130	M	Non-Hodgkin's lymphoma	200,202
124	133	94	117	87	78	102	105	F		
124	250	88	73	11	113	44	94	M	Lymphosarcoma and reticulosarcoma	200
150	228	51	56	6	59	101	75	F		
108	100	86	132	111	57	106	128	M	Hodgkin's disease	201
110	161	126	82	81	46	105	132	F		
103	110	111	107	97	78	93	133	M	Other malignant neoplasm of	202
123	127	96	121	92	79	102	107	F	lymphoid and histiocytic tissue	
93	116	126	100	93	69	73	130	M	Multiple myeloma and	203
101	106	143	119	78	83	91	135	F	immunoproliferative neoplasms	
97	125	119	120	96	71	89	157	M	All leukaemias	204-208
99	136	109	112	91	77	87	135	F		
73	128	123	146	91	74	93	158	M	Lymphoid leukaemia	204
79	166	94	127	75	96	78	154	F		
111	122	115	92	93	72	88	161	M	Myeloid leukaemia	205
111	104	121	94	99	63	98	119	F		
73	70	400	151	88	98	30	187	M	Monocytic leukaemia	206
129	261	254	113	172	239	-	94	F		
248	463	-	-	138	-	94	118	M	Other specified leukaemia	207
-	303	-	-	213	-	-	168	F		
167	119	57	147	155	43	77	109	M	Leukaemia of unspecified cell type	208
125	156	78	162	110	29	77	145	F		
-	390	45	367	-	45	-	477	M	Benign neoplasm of bladder	223.3
-	74	206	64	-	-	-	1061	F		
135	188	111	121	37	43	87	92	M	Benign neoplasm of brain and	225
145	159	136	129	25	62	97	101	F	other parts of nervous system	
53	183	88	151	34	117	91	135	M	Benign neoplasm of pituitary	227.3
97	165	112	96	17	71	142	171	F	gland and craniopharyngeal duct	
-	-	-	-	-	-	-	-	M	Benign neoplasm of pineal gland	227.4
-	-	-	-	-	-	-	-	F		
59	157	-	127	28	109	141	274	M	Carcinoma in situ of digestive	230
59	176	106	190	37	58	153	245	F	organs	

Table 6 Series MB1 no. 25

Table 6 Standardised registration ratios - *continued*

ICD (9th Revision) number	Site description		England	Northern	Yorkshire	Trent	East Anglian	North West Thames	North East Thames	South East Thames
				\multicolumn{7}{c}{Regional health authority of residence}
231	Carcinoma in situ of respiratory system	M F	98 101	20 -	96 209	144 82	99 180	59 117	81 36	170 66
232	Carcinoma in situ of skin	M F	101 100	80 95	166 212	120 117	138 169	68 64	60 46	44 56
233	Carcinoma in situ of breast and genitourinary system	M F	97 101	205 119	655 136	50 91	77 140	35 67	29 80	30 56
233.0	Carcinoma in situ of breast	M F	85 100	- 58	141 86	- 103	- 117	- 88	146 69	- 127
233.1	Carcinoma in situ of cervix uteri	F	101	124	135	91	143	67	81	50
234	Carcinoma in situ of other and unspecified sites	M F	89 85	- -	- -	175 -	- -	- -	- -	- -
235	Neoplasm of uncertain behaviour of digestive and respiratory systems	M F	94 95	82 55	55 79	74 73	132 127	83 115	80 89	62 83
236	Neoplasm of uncertain behaviour of genitourinary organs	M F	85 101	5 928	9 51	51 41	1083 163	59 35	69 42	53 32
237	Neoplasm of uncertain behaviour of endocrine glands and nervous system	M F	93 93	65 64	47 29	83 105	61 51	84 118	166 109	146 180
238	Neoplasm of uncertain behaviour of other and unspecified sites and tissues	M F	82 89	30 60	137 137	69 91	73 89	68 94	90 59	51 73
239.4	Neoplasm of unspecified nature of bladder	M F	31 33	82 64	34 -	52 -	- -	- -	- -	- -
239.6	Neoplasm of unspecified nature of brain	M F	94 101	174 124	94 114	126 111	- -	- -	- -	- -
239.7	Neoplasm of unspecified nature of other parts of nervous system and pituitary gland only	M F	100 85	106 -	88 -	134 217	- 191	- -	- -	- -
630	Hydatidiform mole	F	106	41	114	250	199	74	36	60

Series MB1 no. 25 Table 6

South West Thames	Wessex	Oxford	South Western	West Midlands	Mersey	North Western	Wales		Site description	ICD (9th Revision) number
158	30	153	26	18	244	156	124	M	Carcinoma in situ of respiratory	231
164	162	-	176	-	158	128	84	F	system	
70	36	178	152	147	-	116	87	M	Carcinoma in situ of skin	232
47	38	157	144	126	1	104	107	F		
43	43	61	51	7	45	38	153	M	Carcinoma in situ of breast and	233
60	118	153	65	115	141	116	83	F	genitourinary system	
169	158	222	275	98	-	-	334	M	Carcinoma in situ of breast	233.0
150	148	102	95	90	99	95	92	F		
52	116	160	61	119	146	118	79	F	Carcinoma in situ of cervix uteri	233.1
284	540	-	-	-	357	-	279	M	Carcinoma in situ of other and	234
312	311	464	269	-	-	-	336	F	unspecified sites	
98	255	5	136	47	145	112	188	M	Neoplasm of uncertain behaviour	235
89	300	9	157	32	54	95	173	F	of digestive and respiratory systems	
38	26	-	23	35	-	64	347	M	Neoplasm of uncertain behaviour	236
20	105	2	47	4	28	72	81	F	of genitourinary organs	
125	110	11	102	132	62	47	216	M	Neoplasm of uncertain behaviour	237
167	114	7	126	108	13	42	211	F	of endocrine glands and nervous system	
63	209	42	155	41	75	48	391	M	Neoplasm of uncertain behaviour	238
58	203	47	167	57	71	40	274	F	of other and unspecified sites and tissues	
-	38	-	65	-	-	130	1176	M	Neoplasm of unspecified nature of	239.4
-	120	-	52	-	-	196	1149	F	bladder	
-	213	-	114	219	65	176	190	M	Neoplasm of unspecified nature of	239.6
-	261	-	95	334	72	113	91	F	brain	
-	296	-	260	62	-	331	107	M	Neoplasm of unspecified nature of	239.7
-	513	-	111	-	-	153	330	F	other parts of nervous system and pituitary gland only	
67	257	55	176	28	-	141	-	F	Hydatidiform mole	630

Table 9 Series MB1 no. 25

Table 9 Cancer mortality to incidence ratios: site, sex and regional health authority of residence, 1992

ICD (9th Revision) number	Site description		England and Wales	England	Northern	Yorkshire	Trent	East Anglian	North West Thames	North East Thames
					\multicolumn{6}{l}{Regional health authority of residence}
140-208	All malignant neoplasms	M	0.59	0.59	0.68	0.58	0.59	0.55	0.58	0.64
		F	0.53	0.53	0.59	0.52	0.54	0.51	0.54	0.57
140-208 x 173	All malignant neoplasms excluding 173	M	0.69	0.69	0.78	0.68	0.70	0.70	0.65	0.71
		F	0.61	0.61	0.68	0.61	0.63	0.62	0.59	0.63
140-149	Lip, mouth and pharynx	M	0.48	0.48	0.54	0.53	0.52	0.30	0.43	0.46
		F	0.46	0.47	0.39	0.47	0.48	0.47	0.43	0.43
140	Malignant neoplasm of lip	M	0.12	0.11	0.44	0.10	-	-	0.67	-
		F	0.22	0.25	-	-	1.00	-	-	0.50
141	Malignant neoplasm of tongue	M	0.53	0.53	0.46	0.53	0.46	0.79	0.90	0.54
		F	0.46	0.47	0.41	0.59	0.21	0.46	0.65	0.44
142	Malignant neoplasm of major salivary glands	M	0.47	0.50	0.69	0.28	1.00	-	0.54	0.39
		F	0.37	0.38	0.37	0.09	0.38	0.33	0.31	0.38
143	Malignant neoplasm of gum	M	0.81	0.82	0.75	2.25	0.44	0.33	0.50	0.40
		F	0.44	0.45	1.00	0.50	0.43	-	-	1.00
144	Malignant neoplasm of floor of mouth	M	0.32	0.32	0.34	0.46	0.27	0.20	0.25	0.33
		F	0.41	0.43	0.29	0.57	0.57	-	0.25	0.18
145	Malignant neoplasm of other and unspecified parts of mouth	M	0.45	0.43	0.80	0.48	0.74	1.20	0.05	0.31
		F	0.42	0.43	0.25	0.70	0.38	0.57	0.60	0.46
146	Malignant neoplasm of oropharynx	M	0.49	0.50	0.52	0.80	0.52	0.13	0.50	0.33
		F	0.45	0.47	0.50	0.73	0.58	1.00	0.17	-
147	Malignant neoplasm of nasopharynx	M	0.67	0.70	1.33	0.71	0.56	2.00	0.29	0.67
		F	0.88	0.83	0.57	1.00	0.60	1.00	0.60	0.80
148	Malignant neoplasm of hypopharynx	M	0.42	0.41	0.20	0.55	0.59	0.21	0.30	0.35
		F	0.49	0.48	0.15	0.71	0.69	0.50	0.31	1.00
149	Malignant neoplasm of other and ill-defined sites within the lip, oral cavity and pharynx	M	0.94	0.93	1.11	0.78	1.00	0.67	0.57	2.25
		F	0.77	0.82	0.83	0.67	1.40	-	0.40	0.50
150	Malignant neoplasm of oesophagus	M	0.98	0.97	0.97	1.18	0.87	1.40	0.87	1.02
		F	0.89	0.89	1.02	0.92	0.81	0.95	0.87	0.94
151	Malignant neoplasm of stomach	M	0.79	0.79	0.91	0.78	0.76	0.83	0.71	0.75
		F	0.84	0.85	0.93	0.80	0.89	1.02	0.78	0.88
152	Malignant neoplasm of small intestine, including duodenum	M	0.58	0.58	1.08	0.43	0.43	0.50	1.00	0.69
		F	0.57	0.59	1.00	0.46	1.00	0.67	1.09	0.53
153,154	Colorectal neoplasms	M	0.58	0.59	0.69	0.54	0.60	0.58	0.56	0.58
		F	0.59	0.59	0.62	0.55	0.60	0.57	0.65	0.61
153	Malignant neoplasm of colon	M	0.64	0.65	0.75	0.60	0.68	0.62	0.65	0.67
		F	0.63	0.63	0.67	0.56	0.65	0.61	0.72	0.65
154	Malignant neoplasm of rectum, rectosigmoid junction and anus	M	0.51	0.51	0.62	0.46	0.50	0.53	0.42	0.46
		F	0.50	0.51	0.51	0.53	0.50	0.48	0.50	0.54
155	Malignant neoplasm of liver and intrahepatic bile ducts	M	0.98	1.02	1.78	1.19	0.86	1.79	0.76	0.85
		F	1.19	1.25	1.81	1.21	0.81	2.82	1.08	1.04
156	Malignant neoplasm of gallbladder and extrahepatic bile ducts	M	0.52	0.53	0.23	0.62	0.49	0.21	0.57	0.63
		F	0.66	0.65	0.51	0.53	0.70	0.47	0.57	0.91
157	Malignant neoplasm of pancreas	M	0.97	0.97	0.89	1.01	0.94	0.99	0.93	1.02
		F	0.97	0.97	0.87	0.99	0.97	1.07	0.89	0.90
158	Malignant neoplasm of retroperitoneum and peritoneum	M	0.82	0.87	1.20	1.11	0.83	0.67	0.50	0.89
		F	0.53	0.55	0.50	0.45	1.00	1.25	0.67	0.30

Series MB1 no. 25 Table 9

**England and Wales,
England, Wales, and
regional health authorities
Registered by July 1998**

South East Thames	South West Thames	Wessex	Oxford	South Western	West Midlands	Mersey	North Western	Wales		Site description	ICD (9th Revision) number
0.66	**0.59**	**0.58**	**0.48**	**0.55**	**0.57**	**0.61**	**0.61**	**0.51**	M	All malignant neoplasms	140-208
0.54	**0.55**	**0.49**	**0.44**	**0.51**	**0.53**	**0.55**	**0.53**	**0.47**	F		
0.73	**0.67**	**0.62**	**0.65**	**0.68**	**0.68**	**0.73**	**0.71**	**0.61**	M	All malignant neoplasms	140-208
0.61	**0.62**	**0.52**	**0.55**	**0.61**	**0.63**	**0.68**	**0.62**	**0.54**	F	excluding 173	x 173
0.44	0.48	0.41	0.49	0.54	0.49	0.49	0.47	0.47	M	Lip, mouth and pharynx	140-149
0.61	0.53	0.37	0.44	0.48	0.53	0.44	0.48	0.34	F		
0.20	0.25	0.17	-	0.05	0.50	1.00	0.10	0.19	M	Malignant neoplasm of lip	140
0.17	-	0.43	0.17	-	0.33	-	0.13	-	F		
0.54	0.59	0.47	0.54	0.55	0.38	0.50	0.55	0.61	M	Malignant neoplasm of tongue	141
0.79	0.71	0.16	0.38	0.67	0.60	0.22	0.50	0.43	F		
0.38	0.43	0.52	1.00	0.50	0.72	0.15	0.57	0.28	M	Malignant neoplasm of major	142
0.69	0.46	0.42	0.17	0.43	0.50	0.55	0.33	0.26	F	salivary glands	
0.25	0.50	0.75	-	-	0.22	2.00	3.33	0.75	M	Malignant neoplasm of gum	143
0.43	0.20	0.60	-	1.00	-	0.67	0.56	0.20	F		
0.38	0.07	0.25	0.60	0.33	0.24	0.55	0.29	0.32	M	Malignant neoplasm of floor of	144
0.60	0.67	0.29	1.33	1.00	0.14	0.30	0.67	0.14	F	mouth	
0.23	0.67	0.24	0.23	0.36	0.45	0.45	0.41	0.73	M	Malignant neoplasm of other and	145
0.30	0.64	0.13	0.75	0.27	0.69	0.36	0.35	0.25	F	unspecified parts of mouth	
0.23	0.67	0.35	0.25	0.76	0.95	0.32	0.50	0.36	M	Malignant neoplasm of oropharynx	146
0.38	0.67	2.00	-	0.40	0.70	0.20	0.50	0.31	F		
1.00	0.60	0.70	0.67	1.67	0.47	0.60	0.50	0.30	M	Malignant neoplasm of nasopharynx	147
1.25	0.75	4.00	2.00	0.50	1.20	0.60	0.43	4.00	F		
0.67	0.29	0.50	1.50	0.25	0.54	0.50	0.18	0.56	M	Malignant neoplasm of hypopharynx	148
0.67	0.27	0.71	0.75	0.27	0.41	0.27	0.73	0.67	F		
1.14	0.44	0.38	0.67	2.00	0.83	0.82	1.57	1.00	M	Malignant neoplasm of other and	149
-	0.67	1.00	-	0.43	1.00	1.33	0.83	0.20	F	ill-defined sites within the lip, oral cavity and pharynx	
0.95	0.91	0.84	0.93	1.02	0.94	1.10	0.93	1.14	M	Malignant neoplasm of oesophagus	150
0.87	0.82	0.79	0.85	1.03	0.87	1.01	0.83	0.84	F		
0.79	0.81	0.79	0.81	0.80	0.78	0.76	0.83	0.72	M	Malignant neoplasm of stomach	151
0.83	0.78	0.71	0.68	0.83	0.90	0.92	0.89	0.78	F		
0.57	0.73	0.22	0.78	0.62	0.63	0.78	0.50	0.50	M	Malignant neoplasm of small	152
0.17	0.54	0.39	0.45	0.40	0.60	1.00	0.56	0.40	F	intestine, including duodenum	
0.65	0.58	0.52	0.56	0.59	0.56	0.67	0.62	0.52	M	Colorectal neoplasms	153,154
0.62	0.58	0.52	0.55	0.58	0.58	0.57	0.65	0.53	F		
0.77	0.62	0.54	0.63	0.66	0.56	0.70	0.70	0.59	M	Malignant neoplasm of colon	153
0.69	0.61	0.55	0.60	0.61	0.63	0.62	0.66	0.56	F		
0.51	0.53	0.48	0.47	0.49	0.55	0.63	0.52	0.44	M	Malignant neoplasm of rectum,	154
0.47	0.52	0.45	0.45	0.52	0.50	0.46	0.64	0.46	F	rectosigmoid junction and anus	
0.92	0.96	0.70	1.33	1.06	1.46	0.85	1.05	0.66	M	Malignant neoplasm of liver and	155
1.07	1.31	0.79	1.95	1.10	2.96	1.43	1.02	0.75	F	intrahepatic bile ducts	
0.65	0.86	0.56	0.65	0.39	0.44	0.86	0.72	0.44	M	Malignant neoplasm of gallbladder	156
0.78	0.70	0.49	0.69	0.83	0.61	0.50	0.82	0.80	F	and extrahepatic bile ducts	
0.96	0.95	0.91	1.05	1.03	0.97	1.08	0.96	0.86	M	Malignant neoplasm of pancreas	157
0.98	1.03	0.97	0.91	0.99	1.05	1.06	1.00	0.84	F		
1.00	0.50	0.42	0.50	1.60	0.64	1.60	1.11	0.22	M	Malignant neoplasm of	158
0.56	0.50	0.24	0.60	0.13	0.82	1.00	0.53	0.33	F	retroperitoneum and peritoneum	

Table 9 Series MB1 no. 25

Table 9 Cancer mortality to incidence ratios - *continued*

ICD (9th Revision) number	Site description		England and Wales	England	Northern	Yorkshire	Trent	East Anglian	North West Thames	North East Thames
					Regional health authority of residence					
159	Malignant neoplasm of other and ill-defined sites within the digestive organs and peritoneum	M F	**2.03** **2.11**	2.02 2.08	2.05 3.06	3.00 1.76	3.00 2.36	2.63 4.83	1.55 1.57	1.29 1.33
160	Malignant neoplasm of nasal cavities, middle ear and accessory sinuses	M F	**0.47** **0.36**	0.45 0.35	0.22 0.86	0.32 0.22	0.71 0.50	0.27 0.17	0.27 0.50	0.56 0.50
161	Malignant neoplasm of larynx	M F	**0.42** **0.52**	0.42 0.51	0.39 0.62	0.57 0.46	0.40 0.61	0.35 0.33	0.40 0.32	0.45 0.36
162	Malignant neoplasm of trachea, bronchus and lung	M F	**0.91** **0.89**	0.91 0.89	0.93 0.91	0.91 0.91	0.92 0.92	0.95 0.88	0.85 0.85	0.92 0.89
163	Malignant neoplasm of pleura	M F	**0.53** **0.49**	0.53 0.50	0.54 0.38	0.58 1.13	0.81 0.61	0.62 0.50	0.48 0.31	0.44 0.29
164	Malignant neoplasm of thymus, heart and mediastinum	M F	**0.45** **0.41**	0.45 0.39	0.57 0.33	0.43 0.25	0.67 0.36	0.75 -	0.67 0.29	0.38 -
165	Other malignant neoplasms within the respiratory system and intrathoracic organs	M F	**19.00** **3.20**	18.00 4.00	- -	- -	- -	1.00 -	- -	- -
170	Malignant neoplasm of bone and articular cartilage	M F	**0.56** **0.71**	0.60 0.78	0.80 1.67	1.00 0.63	0.65 0.91	1.60 1.25	0.53 0.39	0.48 0.58
171	Malignant neoplasm of connective and other soft tissue	M F	**0.43** **0.48**	0.44 0.49	0.48 0.53	0.29 0.81	0.77 0.65	0.35 0.50	0.52 0.83	0.67 0.90
172	Malignant melanoma of skin	M F	**0.34** **0.23**	0.34 0.23	0.46 0.31	0.23 0.19	0.31 0.28	0.26 0.21	0.41 0.20	0.37 0.20
173	Other malignant neoplasm of skin	M F	**0.01** **0.01**	0.01 0.01	0.01 0.02	0.01 0.01	0.01 0.01	0.00 0.00	0.01 0.03	0.02 0.02
174	Malignant neoplasm of female breast	F	**0.43**	0.43	0.47	0.42	0.45	0.47	0.39	0.43
175	Malignant neoplasm of male breast	M	**0.48**	0.47	0.70	0.69	1.00	0.56	0.50	0.31
179	Malignant neoplasm of uterus, part unspecified	F	**1.12**	1.23	0.95	0.94	1.00	4.17	26.00	-
180	Malignant neoplasm of cervix uteri	F	**0.48**	0.49	0.42	0.45	0.52	0.45	0.56	0.55
181	Malignant neoplasm of placenta	F	**0.11**	0.11	-	-	-	-	-	-
182	Malignant neoplasm of body of uterus	F	**0.23**	0.23	0.31	0.25	0.27	0.21	0.21	0.20
183	Malignant neoplasm of ovary and other uterine adnexa	F	**0.72**	0.72	0.82	0.73	0.71	0.77	0.74	0.65
184	Malignant neoplasm of other and unspecified female genital organs	F	**0.45**	0.45	0.36	0.35	0.44	0.67	0.34	0.58
185	Malignant neoplasm of prostate	M	**0.56**	0.56	0.69	0.53	0.55	0.48	0.53	0.58
186	Malignant neoplasm of testis	M	**0.09**	0.09	0.14	0.10	0.11	0.12	0.09	0.06
187	Malignant neoplasm of penis and other male genital organs	M	**0.29**	0.29	0.11	0.41	0.19	0.13	0.15	0.29
188	Malignant neoplasm of bladder	M F	**0.41** **0.46**	0.41 0.47	0.48 0.45	0.45 0.59	0.43 0.45	0.75 0.56	0.44 0.46	0.43 0.60
189	Malignant neoplasm of kidney and other and unspecified urinary organs	M F	**0.57** **0.59**	0.58 0.60	0.71 0.62	0.50 0.59	0.59 0.58	0.58 0.50	0.49 0.56	0.60 0.69
190	Malignant neoplasm of eye	M F	**0.23** **0.37**	0.23 0.35	0.22 0.46	0.16 0.44	0.44 0.26	0.43 0.80	0.07 0.50	0.26 0.15

South East Thames	South West Thames	Wessex	Oxford	South Western	West Midlands	Mersey	North Western	Wales		Site description	ICD (9th Revision) number
1.03	1.38	2.06	2.60	4.40	-	4.20	0.97	2.25	M	Malignant neoplasm of other and ill-defined sites within the digestive organs and peritoneum	159
1.03	1.11	1.96	2.75	4.20	59.00	19.00	1.59	2.85	F		
1.08	0.07	0.19	0.33	0.36	0.91	0.20	0.41	0.75	M	Malignant neoplasm of nasal cavities, middle ear and accessory sinuses	160
0.58	0.17	0.25	0.09	0.38	0.24	0.30	0.33	0.47	F		
0.47	0.40	0.40	0.40	0.61	0.35	0.49	0.34	0.40	M	Malignant neoplasm of larynx	161
0.44	0.27	0.42	0.58	0.67	0.59	0.48	0.81	0.57	F		
0.91	0.89	0.90	0.88	0.97	0.92	0.89	0.90	0.82	M	Malignant neoplasm of trachea, bronchus and lung	162
0.86	0.91	0.78	0.88	0.89	0.93	0.92	0.90	0.86	F		
0.37	0.49	0.67	0.66	0.47	0.40	0.53	0.61	0.48	M	Malignant neoplasm of pleura	163
0.50	0.43	0.44	0.70	0.14	0.45	0.56	0.67	0.30	F		
0.40	-	0.36	0.25	0.50	0.36	1.00	0.50	0.44	M	Malignant neoplasm of thymus, heart and mediastinum	164
0.50	0.50	-	-	4.00	0.60	0.67	0.25	0.50	F		
-	-	-	-	1.00	-	-	-	-	M	Other malignant neoplasms within the respiratory system and intrathoracic organs	165
-	-	0.33	-	-	-	-	-	-	F		
0.82	0.63	0.30	0.67	0.53	0.60	0.50	0.42	0.23	M	Malignant neoplasm of bone and articular cartilage	170
1.00	0.67	0.59	1.00	0.33	1.25	2.00	0.56	0.26	F		
0.16	0.81	0.28	0.40	0.44	0.48	0.42	0.58	0.34	M	Malignant neoplasm of connective and other soft tissue	171
0.21	0.78	0.27	0.70	0.62	0.30	0.60	0.59	0.33	F		
0.52	0.35	0.31	0.28	0.31	0.33	0.37	0.35	0.40	M	Malignant melanoma of skin	172
0.32	0.29	0.16	0.15	0.23	0.29	0.29	0.20	0.16	F		
0.03	0.03	0.04	0.01	0.01	0.01	0.02	0.02	0.01	M	Other malignant neoplasm of skin	173
0.01	0.01	0.04	0.01	0.01	0.01	0.01	0.01	0.01	F		
0.44	0.47	0.40	0.39	0.43	0.44	0.48	0.40	0.37	F	Malignant neoplasm of female breast	174
0.25	0.56	0.35	0.40	0.29	0.45	0.33	0.50	0.58	M	Malignant neoplasm of male breast	175
0.46	35.00	0.70	2.83	1.27	1.02	4.25	0.94	0.46	F	Malignant neoplasm of uterus, part unspecified	179
0.55	0.42	0.39	0.40	0.66	0.52	0.50	0.39	0.44	F	Malignant neoplasm of cervix uteri	180
-	-	-	-	-	0.50	-	-	-	F	Malignant neoplasm of placenta	181
0.23	0.26	0.23	0.16	0.24	0.23	0.18	0.22	0.28	F	Malignant neoplasm of body of uterus	182
0.61	0.76	0.58	0.71	0.85	0.79	0.76	0.73	0.63	F	Malignant neoplasm of ovary and other uterine adnexa	183
0.26	0.46	0.47	0.51	0.57	0.47	0.74	0.59	0.33	F	Malignant neoplasm of other and unspecified female genital organs	184
0.66	0.55	0.51	0.51	0.55	0.55	0.58	0.58	0.49	M	Malignant neoplasm of prostate	185
0.08	0.04	0.06	0.10	0.13	0.06	0.08	0.10	0.08	M	Malignant neoplasm of testis	186
0.17	0.47	0.29	0.13	0.59	0.40	0.82	0.10	0.33	M	Malignant neoplasm of penis and other male genital organs	187
0.49	0.36	0.32	0.36	0.38	0.34	0.40	0.41	0.31	M	Malignant neoplasm of bladder	188
0.54	0.41	0.38	0.34	0.47	0.45	0.47	0.47	0.38	F		
0.58	0.44	0.51	0.67	0.51	0.63	0.63	0.72	0.46	M	Malignant neoplasm of kidney and other and unspecified urinary organs	189
0.73	0.67	0.48	0.67	0.46	0.53	0.72	0.65	0.57	F		
0.26	0.33	0.04	0.33	0.36	0.29	0.33	0.11	0.30	M	Malignant neoplasm of eye	190
0.42	0.18	0.47	0.05	0.50	0.55	0.40	0.13	0.64	F		

Table 9 Cancer mortality to incidence ratios - *continued*

ICD (9th Revision) number	Site description		England and Wales	England	Northern	Yorkshire	Trent	East Anglian	North West Thames	North East Thames
					\multicolumn{6}{l}{Regional health authority of residence}
191	Malignant neoplasm of brain	M	**0.82**	0.84	0.78	0.78	0.87	0.83	0.89	0.95
		F	**0.76**	0.77	0.67	0.75	0.80	0.81	0.83	0.77
192	Malignant neoplasm of other and unspecified parts of nervous system	M	**0.37**	0.37	3.00	0.17	0.33	-	0.40	3.00
		F	**0.34**	0.36	0.33	0.33	0.09	-	1.00	1.00
193	Malignant neoplasm of thyroid gland	M	**0.35**	0.38	0.44	0.33	0.53	0.14	0.33	0.25
		F	**0.33**	0.33	0.32	0.35	0.27	0.38	0.51	0.36
194	Malignant neoplasm of other endocrine glands and related structures	M	**0.57**	0.55	0.83	0.67	1.50	0.20	0.36	0.60
		F	**0.51**	0.53	2.00	0.50	0.71	0.33	0.50	1.00
195	Malignant neoplasm of other and ill-defined sites	M	**1.04**	1.34	2.00	1.60	1.58	2.67	2.50	1.60
		F	**0.92**	1.07	1.10	0.85	0.90	1.67	4.50	0.64
196	Secondary and unspecified malignant neoplasm of lymph nodes	M	-	-	-	-	-	-	-	-
		F	-	-	-	-	-	-	-	-
197	Secondary malignant neoplasm of respiratory and digestive systems	M	-	-	-	-	-	-	-	-
		F	-	-	-	-	-	-	-	-
198	Secondary malignant neoplasm of other specified sites	M	-	-	-	-	-	-	-	-
		F	-	-	-	-	-	-	-	-
199	Malignant neoplasm without specification of site	M	**1.74**	1.74	1.91	1.83	2.17	1.64	1.72	1.78
		F	**1.67**	1.67	1.76	1.78	1.92	1.30	1.85	1.98
200,202	Non-Hodgkin's lymphoma	M	**0.53**	0.54	0.63	0.45	0.72	0.52	0.50	0.48
		F	**0.58**	0.58	0.70	0.57	0.75	0.55	0.53	0.60
200	Lymphosarcoma and reticulosarcoma	M	**0.23**	0.23	0.47	0.16	0.14	0.06	0.20	0.21
		F	**0.29**	0.30	0.33	0.16	0.33	0.33	0.50	0.14
201	Hodgkin's disease	M	**0.33**	0.32	0.28	0.41	0.49	0.33	0.33	0.37
		F	**0.31**	0.31	0.39	0.29	0.40	0.37	0.24	0.42
202	Other malignant neoplasm of lymphoid and histiocytic tissue	M	**0.55**	0.57	0.64	0.49	0.78	0.57	0.52	0.51
		F	**0.59**	0.60	0.73	0.62	0.77	0.57	0.54	0.63
203	Multiple myeloma and immunoproliferative neoplasms	M	**0.71**	0.72	0.70	0.58	0.58	0.69	0.53	0.76
		F	**0.76**	0.78	1.02	0.62	0.94	0.70	0.97	0.75
204-208	All leukaemias	M	**0.66**	0.68	0.76	0.78	0.65	0.74	0.71	0.74
		F	**0.68**	0.69	0.86	0.64	0.72	0.72	0.72	0.64
204	Lymphoid leukaemia	M	**0.54**	0.55	0.84	0.69	0.41	0.64	0.69	0.64
		F	**0.51**	0.52	0.49	0.49	0.51	0.67	0.45	0.58
205	Myeloid leukaemia	M	**0.78**	0.81	0.69	0.77	0.90	0.79	0.77	0.86
		F	**0.82**	0.83	1.15	0.72	0.82	0.76	0.88	0.65
206	Monocytic leukaemia	M	**0.62**	0.68	-	2.00	0.50	0.50	1.00	0.67
		F	**0.53**	0.53	0.40	-	1.00	-	1.00	1.00
207	Other specified leukaemia	M	**0.86**	0.85	-	-	0.33	-	-	-
		F	**0.80**	0.89	1.00	1.00	-	-	-	-
208	Leukaemia of unspecified cell type	M	**0.61**	0.62	0.67	2.00	0.78	2.00	0.43	0.64
		F	**0.74**	0.77	1.33	0.57	2.00	0.50	0.65	1.00

Series MB1 no. 25 Table 9

South East Thames	South West Thames	Wessex	Oxford	South Western	West Midlands	Mersey	North Western	Wales		Site description	ICD (9th Revision) number
0.96	0.75	0.88	0.81	0.73	0.81	0.97	0.82	0.60	M	Malignant neoplasm of brain	191
0.66	0.90	0.71	0.53	0.81	0.81	0.84	0.86	0.65	F		
0.57	0.50	0.10	0.14	0.33	0.50	-	-	0.38	M	Malignant neoplasm of other and unspecified parts of nervous system	192
0.33	-	1.50	0.60	0.25	-	0.17	0.33	0.13	F		
0.38	0.22	0.44	0.23	0.48	0.40	0.50	0.36	0.06	M	Malignant neoplasm of thyroid gland	193
0.35	0.40	0.31	0.30	0.33	0.28	0.36	0.23	0.28	F		
1.33	0.20	0.25	0.60	0.45	0.40	0.80	0.60	0.80	M	Malignant neoplasm of other endocrine glands and related structures	194
0.53	0.50	0.29	1.25	0.11	0.32	0.50	0.75	0.29	F		
0.42	2.75	0.50	1.25	0.21	-	1.75	4.00	0.27	M	Malignant neoplasm of other and ill-defined sites	195
0.83	0.85	0.48	0.91	0.63	11.00	1.50	2.60	0.28	F		
-	-	-	-	-	-	-	-	-	M	Secondary and unspecified malignant neoplasm of lymph nodes	196
-	-	-	-	-	-	-	-	-	F		
-	-	-	-	-	-	-	-	-	M	Secondary malignant neoplasm of respiratory and digestive systems	197
-	-	-	-	-	-	-	-	-	F		
-	-	-	-	-	-	-	-	-	M	Secondary malignant neoplasm of other specified sites	198
-	-	-	-	-	-	-	-	-	F		
2.28	2.15	2.40	1.04	2.14	0.82	5.20	2.04	1.79	M	Malignant neoplasm without specification of site	199
2.04	2.07	2.04	1.16	1.81	0.87	4.15	1.85	1.72	F		
0.64	0.47	0.46	0.50	0.58	0.56	0.54	0.53	0.36	M	Non-Hodgkin's lymphoma	200,202
0.58	0.48	0.42	0.65	0.58	0.62	0.75	0.52	0.52	F		
0.62	0.05	0.10	0.36	0.29	1.33	0.07	0.78	0.27	M	Lymphosarcoma and reticulosarcoma	200
0.43	0.13	0.12	0.50	0.86	6.00	-	0.21	0.13	F		
0.27	0.32	0.26	0.34	0.16	0.36	0.37	0.25	0.40	M	Hodgkin's disease	201
0.19	0.15	0.13	0.34	0.22	0.38	0.45	0.50	0.26	F		
0.64	0.51	0.52	0.51	0.59	0.55	0.60	0.52	0.36	M	Other malignant neoplasm of lymphoid and histiocytic tissue	202
0.59	0.50	0.46	0.65	0.57	0.60	0.79	0.54	0.54	F		
0.85	0.80	0.68	0.81	0.69	0.83	0.94	0.78	0.64	M	Multiple myeloma and immunoproliferative neoplasms	203
0.68	0.83	0.79	0.49	0.69	0.94	1.00	0.73	0.58	F		
0.55	0.69	0.57	0.61	0.60	0.77	0.84	0.68	0.45	M	All leukaemias	204-208
0.78	0.66	0.55	0.43	0.63	0.80	0.84	0.80	0.54	F		
0.40	0.71	0.40	0.37	0.40	0.62	0.73	0.59	0.45	M	Lymphoid leukaemia	204
0.62	0.54	0.33	0.43	0.34	0.64	0.66	0.72	0.40	F		
0.66	0.73	0.75	0.89	0.90	0.95	0.96	0.76	0.46	M	Myeloid leukaemia	205
0.86	0.80	0.84	0.46	0.96	0.92	1.14	0.78	0.71	F		
0.83	1.00	1.00	0.25	1.00	0.50	1.00	-	0.20	M	Monocytic leukaemia	206
-	0.67	0.33	-	0.67	0.50	-	-	0.50	F		
-	1.00	0.25	-	-	1.50	-	1.00	1.00	M	Other specified leukaemia	207
0.50	-	-	-	-	1.00	-	-	-	F		
0.70	0.38	0.58	0.50	0.47	0.54	0.67	0.78	0.40	M	Leukaemia of unspecified cell type	208
1.20	0.17	0.67	0.40	0.44	0.63	1.00	1.22	0.46	F		

Table 10 Series MB1 no. 25

Table 10 Directly age standardised* registration rates per 100,000 population: site and sex, 1983 to 1992

England and Wales

ICD (9th Revision) number	Site description	Sex	1983	1984	1985	1986	1987	1988	1989	1990	1991	1992
	All registrations	M	435.1	435.0	459.8	449.0	454.7	472.6	465.0	463.6	471.7	486.5
		F	364.6	375.2	412.3	414.8	435.2	451.5	451.6	463.4	468.1	478.3
140-208	All malignant neoplasms	M	421.1	420.7	445.3	434.9	440.6	456.9	449.0	447.3	454.6	467.3
		F	316.5	318.7	339.6	335.8	348.1	357.4	359.6	358.1	367.4	378.1
140-208 x 173	All malignant neoplasms excluding 173	M	372.0	370.7	388.4	374.8	381.2	391.9	384.5	384.8	388.6	397.6
		F	284.6	286.8	305.0	297.9	309.0	314.7	316.7	316.9	324.7	331.0
140-149	Lip, mouth and pharynx	M	8.0	7.9	8.0	7.9	7.9	8.4	7.9	8.1	8.5	9.0
		F	3.6	3.7	3.6	3.3	3.6	3.6	3.5	3.7	3.6	4.0
140	Malignant neoplasm of lip	M	0.9	0.8	0.8	0.9	0.8	0.8	0.7	0.6	0.5	0.7
		F	0.1	0.2	0.2	0.1	0.2	0.2	0.2	0.1	0.1	0.2
141	Malignant neoplasm of tongue	M	1.4	1.5	1.5	1.5	1.6	1.7	1.7	1.6	1.8	1.8
		F	0.8	0.8	0.8	0.7	0.7	0.8	0.7	0.8	0.8	0.9
142	Malignant neoplasm of major salivary glands	M	0.8	0.7	0.8	0.6	0.8	0.8	0.7	0.7	0.8	0.9
		F	0.5	0.6	0.5	0.6	0.5	0.5	0.5	0.5	0.5	0.6
143	Malignant neoplasm of gum	M	0.3	0.4	0.2	0.3	0.2	0.3	0.2	0.3	0.3	0.3
		F	0.2	0.2	0.2	0.2	0.2	0.2	0.2	0.2	0.1	0.2
144	Malignant neoplasm of floor of mouth	M	0.9	0.9	0.9	1.0	0.9	1.0	0.9	0.8	1.0	0.9
		F	0.2	0.2	0.3	0.2	0.3	0.3	0.3	0.3	0.2	0.3
145	Malignant neoplasm of other and unspecified parts of mouth	M	0.8	0.8	0.9	0.9	0.8	1.0	1.0	1.0	1.1	1.2
		F	0.5	0.4	0.5	0.5	0.5	0.5	0.5	0.5	0.5	0.5
146	Malignant neoplasm of oropharynx	M	1.0	0.9	0.9	0.9	1.0	1.0	1.0	1.2	1.1	1.2
		F	0.3	0.4	0.3	0.4	0.4	0.3	0.4	0.4	0.4	0.4
147	Malignant neoplasm of nasopharynx	M	0.6	0.6	0.5	0.5	0.6	0.6	0.5	0.4	0.5	0.6
		F	0.2	0.2	0.3	0.2	0.3	0.2	0.2	0.2	0.2	0.2
148	Malignant neoplasm of hypopharynx	M	1.0	0.9	1.1	0.9	0.9	0.9	0.9	0.9	0.9	1.0
		F	0.5	0.5	0.5	0.4	0.4	0.4	0.3	0.4	0.4	0.5
149	Malignant neoplasm of other and ill-defined sites within the lip, oral cavity and pharynx	M	0.4	0.4	0.4	0.4	0.4	0.4	0.4	0.5	0.4	0.4
		F	0.2	0.2	0.2	0.1	0.2	0.2	0.2	0.2	0.2	0.2
150	Malignant neoplasm of oesophagus	M	9.2	9.6	10.1	10.5	11.0	11.5	11.2	11.6	11.8	12.4
		F	4.9	5.0	5.0	4.9	5.2	5.3	5.5	5.4	5.5	5.8
151	Malignant neoplasm of stomach	M	28.8	27.2	27.7	27.0	25.4	25.8	24.8	23.5	22.7	22.9
		F	11.8	11.6	11.5	10.5	11.2	10.2	10.1	9.3	9.2	8.9
152	Malignant neoplasm of small intestine, including duodenum	M	0.7	0.7	0.8	0.7	0.7	0.7	0.7	0.7	0.9	0.9
		F	0.5	0.5	0.5	0.5	0.5	0.5	0.6	0.5	0.5	0.6
153,154	Colorectal neoplasms	M	49.3	49.2	50.8	48.6	50.4	51.4	51.8	52.0	51.7	54.5
		F	36.1	35.2	36.3	35.8	36.0	36.4	36.1	35.5	35.8	37.3
153	Malignant neoplasm of colon	M	27.5	27.3	28.0	27.2	28.8	29.2	29.8	29.8	29.8	31.1
		F	23.3	22.8	23.8	23.6	23.8	24.1	23.7	23.6	24.0	24.9
154	Malignant neoplasm of rectum, rectosigmoid junction and anus	M	21.8	21.9	22.8	21.4	21.5	22.2	22.0	22.2	21.9	23.4
		F	12.7	12.4	12.6	12.2	12.2	12.3	12.4	11.9	11.9	12.4
155	Malignant neoplasm of liver and intrahepatic bile ducts	M	2.5	2.4	2.7	2.6	2.9	2.9	2.8	2.9	3.1	3.4
		F	1.2	1.2	1.3	1.1	1.4	1.4	1.3	1.3	1.5	1.5
156	Malignant neoplasm of gallbladder and extrahepatic bile ducts	M	1.9	1.8	1.9	1.9	1.7	2.1	1.8	1.8	2.1	2.1
		F	2.0	2.1	2.1	1.9	2.0	1.8	2.0	1.9	1.8	1.9
157	Malignant neoplasm of pancreas	M	12.0	11.9	12.0	11.5	11.3	11.5	11.0	11.3	11.2	11.1
		F	7.7	7.6	8.1	7.8	8.2	7.9	8.3	8.0	8.0	7.9
158	Malignant neoplasm of retroperitoneum and peritoneum	M	0.6	0.6	0.5	0.5	0.5	0.6	0.5	0.5	0.5	0.4
		F	0.5	0.5	0.5	0.5	0.4	0.4	0.3	0.4	0.3	0.4

* Using the European standard population

Table 10 Directly age standardised* rates - *continued* **England and Wales**

ICD (9th Revision) number	Site description		1983	1984	1985	1986	1987	1988	1989	1990	1991	1992
159	Malignant neoplasm of other and ill-defined sites within the digestive organs and peritoneum	M F	0.7 0.5	0.6 0.5	0.8 0.6	0.7 0.5	0.7 0.6	0.9 0.6	0.9 0.7	0.9 0.6	0.8 0.5	0.9 0.7
160	Malignant neoplasm of nasal cavities, middle ear and accessory sinuses	M F	0.8 0.5	0.8 0.5	0.8 0.5	0.9 0.4	0.9 0.4	0.9 0.4	0.9 0.5	0.9 0.5	0.8 0.4	0.9 0.5
161	Malignant neoplasm of larynx	M F	5.8 0.9	6.2 1.0	6.3 1.0	6.2 1.1	6.2 1.3	6.8 1.2	6.4 1.1	6.3 1.2	6.2 1.1	6.4 1.1
162	Malignant neoplasm of trachea, bronchus and lung	M F	106.9 29.3	103.4 30.0	107.8 32.0	100.2 31.2	98.1 32.9	98.1 33.5	94.0 32.9	91.6 32.7	90.3 33.1	89.7 33.9
163	Malignant neoplasm of pleura	M F	1.6 0.3	1.8 0.3	1.9 0.3	2.3 0.3	2.3 0.4	2.4 0.4	2.9 0.5	3.0 0.6	3.4 0.6	3.5 0.6
164	Malignant neoplasm of thymus, heart and mediastinum	M F	0.3 0.2	0.3 0.2	0.4 0.2	0.3 0.2	0.3 0.2	0.3 0.2	0.4 0.2	0.4 0.2	0.4 0.2	0.4 0.2
165	Other malignant neoplasms within the respiratory system and intrathoracic organs	M F	0.0 0.0	0.0 0.0	0.0 0.0	0.0 0.0	0.0 -	- 0.0	0.0 0.0	0.0 0.0	0.0 0.0	0.0 0.0
170	Malignant neoplasm of bone and articular cartilage	M F	1.0 0.6	0.9 0.7	0.9 0.7	0.9 0.8	1.0 0.7	1.1 0.6	0.9 0.7	1.0 0.7	0.9 0.6	0.9 0.6
171	Malignant neoplasm of connective and other soft tissue	M F	1.8 1.4	1.7 1.4	1.9 1.5	1.8 1.5	1.9 1.6	2.3 1.7	2.2 1.6	2.2 1.7	2.4 1.8	2.6 2.0
172	Malignant melanoma of skin	M F	3.6 6.1	3.5 6.1	4.6 7.6	4.7 7.4	5.3 8.4	6.3 9.0	5.7 8.4	6.0 7.5	5.9 7.7	6.4 8.4
173	Other malignant neoplasm of skin	M F	49.1 31.9	49.9 31.9	56.9 34.6	60.2 37.8	59.4 39.1	64.9 42.7	64.5 42.9	62.5 41.2	66.0 42.6	69.8 47.0
174	Malignant neoplasm of female breast	F	78.0	78.6	85.9	85.3	88.3	90.0	95.3	98.5	105.4	107.4
175	Malignant neoplasm of male breast	M	0.7	0.8	0.8	0.7	0.8	0.7	0.8	0.8	0.8	0.7
179	Malignant neoplasm of uterus, part unspecified	F	1.5	1.3	1.2	1.2	1.2	1.3	1.2	1.0	1.0	1.3
180	Malignant neoplasm of cervix uteri	F	14.8	15.4	16.4	16.1	15.8	16.6	15.1	15.4	13.0	11.9
181	Malignant neoplasm of placenta	F	0.1	0.0	0.0	0.1	0.0	0.0	0.0	0.0	0.0	0.0
182	Malignant neoplasm of body of uterus	F	11.9	12.0	12.4	12.0	11.7	12.3	12.2	12.3	12.6	12.4
183	Malignant neoplasm of ovary and other uterine adnexa	F	16.3	16.2	17.0	16.6	17.4	17.2	17.1	16.7	17.6	17.5
184	Malignant neoplasm of other and unspecified female genital organs	F	2.7	3.0	2.9	2.8	2.8	2.9	2.8	2.8	2.5	2.9
185	Malignant neoplasm of prostate	M	39.3	39.8	42.4	42.3	43.6	45.8	46.0	47.4	50.1	54.3
186	Malignant neoplasm of testis	M	3.7	4.2	4.3	4.6	4.7	5.0	5.1	4.8	5.2	5.2
187	Malignant neoplasm of penis and other male genital organs	M	1.4	1.3	1.5	1.3	1.4	1.4	1.4	1.4	1.4	1.4

* Using the European standard population

Table 10 Directly age standardised* rates - continued

England and Wales

ICD (9th Revision) number	Site description		1983	1984	1985	1986	1987	1988	1989	1990	1991	1992
188	Malignant neoplasm of bladder	M F	28.4 7.6	28.8 7.6	30.2 8.2	29.1 8.3	29.9 8.1	31.5 8.7	30.5 8.9	30.8 8.2	30.6 8.2	30.7 8.7
189	Malignant neoplasm of kidney and other and unspecified urinary organs	M F	8.2 3.8	8.4 3.9	8.6 4.0	8.3 4.0	8.9 4.2	9.2 4.4	10.1 4.5	10.0 4.7	9.9 4.9	10.5 4.9
190	Malignant neoplasm of eye	M F	0.8 0.6	0.8 0.7	0.8 0.6	0.9 0.6	0.8 0.8	0.9 0.7	0.8 0.7	1.0 1.0	0.9 0.8	0.9 0.7
191	Malignant neoplasm of brain	M F	6.5 4.4	6.7 4.6	7.0 4.7	6.7 4.5	7.2 4.6	7.2 5.0	7.4 5.0	7.3 4.8	7.5 4.9	8.0 5.4
192	Malignant neoplasm of other and unspecified parts of nervous system	M F	0.2 0.3	0.3 0.3	0.3 0.3	0.3 0.2	0.3 0.2	0.3 0.4	0.3 0.3	0.4 0.4	0.3 0.4	0.3 0.3
193	Malignant neoplasm of thyroid gland	M F	0.9 2.1	0.9 2.0	1.0 1.8	0.9 2.0	0.9 2.0	0.9 2.1	1.0 2.1	0.9 2.2	0.9 2.3	0.9 2.4
194	Malignant neoplasm of other endocrine glands and related structures	M F	0.3 0.3	0.4 0.4	0.4 0.3	0.3 0.4	0.5 0.4	0.6 0.6	0.5 0.5	0.6 0.5	0.7 0.6	0.5 0.4
195	Malignant neoplasm of other and ill-defined sites	M F	1.0 1.0	1.0 0.9	0.7 0.7	0.6 0.7	0.8 0.9	0.7 0.9	0.7 0.6	0.5 0.6	0.6 0.7	0.5 0.7
196	Secondary and unspecified malignant neoplasm of lymph nodes	M F	1.0 0.6	1.0 0.7	1.2 0.8	1.0 0.7	1.0 0.6	1.0 0.7	1.0 0.6	1.1 0.7	1.1 0.8	0.9 0.7
197	Secondary malignant neoplasm of respiratory and digestive systems	M F	5.0 3.9	6.0 4.2	6.3 4.6	6.9 4.9	5.9 4.1	6.6 4.6	6.3 4.6	6.5 4.5	6.6 4.7	6.4 4.8
198	Secondary malignant neoplasm of other specified sites	M F	3.2 2.3	3.7 2.7	3.7 2.7	3.3 2.5	2.8 2.0	2.8 2.0	3.2 2.1	3.6 2.6	3.3 2.5	3.1 2.5
199	Malignant neoplasm without specification of site	M F	8.9 6.8	8.8 6.3	9.5 7.2	9.3 6.6	12.7 9.0	11.5 8.2	11.6 8.1	11.2 8.0	11.8 8.7	11.6 8.3
200,202	Non-Hodgkin's lymphoma	M F	9.1 6.3	9.6 6.4	10.4 7.3	10.8 7.3	11.8 7.9	12.5 8.7	12.5 8.7	13.2 8.4	13.6 9.2	14.0 9.4
200	Lymphosarcoma and reticulosarcoma	M F	2.0 1.3	1.8 1.1	1.3 1.0	1.3 0.8	1.2 0.6	1.0 0.6	1.0 0.7	1.0 0.6	1.1 0.7	1.0 0.5
201	Hodgkin's disease	M F	3.0 1.8	2.8 1.9	3.1 2.0	3.0 1.8	2.8 1.8	2.6 1.9	2.7 1.7	2.8 1.8	2.5 1.6	2.8 1.9
202	Other malignant neoplasm of lymphoid and histiocytic tissue	M F	7.1 5.0	7.8 5.4	9.1 6.4	9.5 6.5	10.7 7.2	11.5 8.1	11.5 7.9	12.3 7.8	12.5 8.5	13.0 8.9
203	Multiple myeloma and immunoproliferative neoplasms	M F	4.8 3.0	4.6 3.2	5.3 3.6	5.0 3.2	5.3 3.6	5.5 3.5	5.1 3.5	5.1 3.5	5.4 3.4	5.5 3.5
204-208	All leukaemias	M F	10.3 6.7	10.2 6.4	11.1 6.8	10.0 6.3	10.5 6.8	11.3 6.9	10.9 6.8	10.9 6.6	11.5 6.5	10.9 6.6
204	Lymphoid leukaemia	M F	5.1 2.8	4.9 2.8	5.2 2.9	4.7 2.6	4.9 2.8	5.6 3.0	5.1 2.7	5.3 2.7	5.3 2.8	5.0 2.7
205	Myeloid leukaemia	M F	4.0 3.1	4.4 3.0	4.8 3.3	4.5 3.2	4.6 3.4	4.7 3.3	4.9 3.4	4.8 3.3	5.2 3.3	5.1 3.5
206	Monocytic leukaemia	M F	0.3 0.2	0.2 0.1	0.2 0.1	0.2 0.1	0.2 0.1	0.2 0.1	0.1 0.1	0.1 0.1	0.2 0.0	0.2 0.1
207	Other specified leukaemia	M F	0.1 0.1	0.1 0.1	0.1 0.1	0.1 0.1	0.1 0.1	0.1 0.1	0.1 0.0	0.1 0.0	0.1 0.0	0.1 0.0
208	Leukaemia of unspecified cell type	M F	0.8 0.6	0.6 0.4	0.8 0.5	0.6 0.3	0.6 0.4	0.8 0.4	0.6 0.5	0.6 0.4	0.7 0.3	0.6 0.4

* Using the European standard population

Table 10 Directly age standardised* rates - *continued* England and Wales

ICD (9th Revision) number	Site description		1983	1984	1985	1986	1987	1988	1989	1990	1991	1992
223.3	Benign neoplasm of bladder	M F	0.3 0.1	0.2 0.1	0.2 0.1	0.1 0.1	0.1 0.1	0.1 0.1	0.1 0.0	0.1 0.1	0.1 0.1	0.2 0.1
225	Benign neoplasm of brain and other parts of nervous system	M F	1.3 2.4	1.5 2.3	1.4 2.4	1.4 2.3	1.5 2.3	1.5 2.4	1.4 2.2	1.4 2.3	1.5 2.4	1.8 3.0
227.3	Benign neoplasm of pituitary gland and craniopharyngeal duct	M F	0.8 0.8	0.8 0.7	0.8 0.7	0.8 0.7	0.8 0.7	0.9 0.8	0.9 0.9	1.0 0.9	1.0 1.0	1.0 0.9
227.4	Benign neoplasm of pineal gland	M F	0.0 0.0	- -	0.0 -	- 0.0	- 0.0	- 0.0	- -	0.0 0.0	- -	- -
230	Carcinoma in situ of digestive organs	M F	0.4 0.2	0.3 0.3	0.4 0.3	0.5 0.3	0.4 0.2	0.4 0.2	0.4 0.2	0.4 0.2	0.4 0.3	0.5 0.3
231	Carcinoma in situ of respiratory system	M F	0.4 0.1	0.5 0.1	0.5 0.1	0.5 0.2	0.4 0.1	0.5 0.1	0.5 0.1	0.6 0.1	0.5 0.1	0.6 0.1
232	Carcinoma in situ of skin	M F	2.0 2.5	1.9 2.6	2.3 3.1	2.6 3.3	2.8 3.9	3.5 5.0	3.5 5.4	3.4 5.5	3.7 5.3	4.1 6.1
233	Carcinoma in situ of breast and genitourinary system	M F	0.7 28.9	0.8 41.8	1.1 57.6	1.2 63.1	1.1 71.1	1.3 76.0	1.6 73.2	1.6 86.2	1.8 81.3	2.2 78.5
233.0	Carcinoma in situ of breast	M F	0.0 2.3	0.0 2.8	0.0 3.6	0.0 3.6	0.0 3.6	0.0 3.8	0.0 3.8	0.0 4.6	0.0 6.3	0.0 6.9
233.1	Carcinoma in situ of cervix uteri	F	25.9	38.3	53.1	58.6	66.4	71.1	67.6	80.1	73.3	69.6
234	Carcinoma in situ of other and unspecified sites	M F	0.0 0.0	0.0 0.0	0.0 0.0	0.0 0.0	0.0 0.0	0.0 0.0	0.0 0.1	0.1 0.0	0.1 0.0	0.0 0.0
235	Neoplasm of uncertain behaviour of digestive and respiratory systems	M F	1.4 1.5	1.5 1.4	1.5 1.5	1.3 1.5	1.4 1.4	1.5 1.4	1.7 1.4	1.5 1.4	1.7 1.6	1.8 1.5
236	Neoplasm of uncertain behaviour of genitourinary organs	M F	2.0 7.5	2.1 2.9	1.5 2.6	1.1 3.3	0.9 2.8	1.0 3.0	1.1 3.5	1.3 3.9	1.1 3.8	1.1 4.4
237	Neoplasm of uncertain behaviour of endocrine glands and nervous system	M F	0.8 0.8	0.8 0.8	0.8 0.6	0.8 0.7	0.7 0.7	0.9 0.9	0.8 0.8	1.0 0.8	1.2 1.1	1.5 1.1
238	Neoplasm of uncertain behaviour of other and unspecified sites and tissues	M F	2.7 1.6	2.7 1.7	2.8 2.0	2.7 2.0	2.9 2.2	3.1 2.6	3.2 2.7	2.9 2.4	3.4 2.5	3.7 2.7
239.4	Neoplasm of unspecified nature of bladder	M F	0.2 0.0	0.2 0.0	0.1 0.0	0.1 0.0	0.1 0.0	0.1 0.0	0.1 0.0	0.1 0.0	0.1 0.0	0.1 0.1
239.6	Neoplasm of unspecified nature of brain	M F	0.8 0.6	0.8 0.7	0.8 0.7	0.8 0.7	0.8 0.6	0.8 0.7	0.7 0.7	0.7 0.6	0.4 0.5	0.5 0.4
239.7	Neoplasm of unspecified nature of other parts of nervous system and pituitary gland only	M F	0.1 0.1	0.1 0.1	0.2 0.1	0.1 0.1	0.1 0.1	0.2 0.1	0.1 0.1	0.1 0.1	0.0 0.1	0.1 0.1
630	Hydatidiform mole	F	0.9	0.9	0.7	0.7	0.8	0.6	0.7	0.9	0.8	0.9

* Using the European standard population

Appendix 1 Guidance notes and definitions

Cancer

For the purposes of the national cancer registration scheme the term 'cancer' includes all malignant neoplasms and the reticuloses, that is conditions listed under site code numbers 140 to 208 of the ICD, Ninth Revision.[1] In addition, all carcinoma in situ and neoplasms of uncertain behaviour are registered. Benign neoplasms and neoplasms of unspecified nature of bladder and brain, including the pineal and pituitary glands, are also registered, together with hydatidiform mole.

It should be noted that some cancer registries are not always able to collect complete information about benign, uncertain and unspecified neoplasms and therefore these registration rates are almost certainly underestimates of the true incidence. In particular this should be noted when interpreting regional differences (**Tables 4** to **6**).

Validity

A brief history of cancer registration in England and Wales is presented above. The essential features of the current system have remained unchanged for over 25 years. The main flows of information to and from the regional registries and ONS (including the NHSCR) are illustrated in Figure A on page 4. Some aspects of the system which are relevant to the interpretation of the data have been discussed in considerable detail by Swerdlow.[2] These include completeness of registration; accuracy; completeness of flagging at NHSCR; geographic coverage; late registrations, deletions and amendments; duplicate and multiple registrations; registrations from death certificates; clinical and pathological definitions and diagnoses; changes in coding systems; changes in definition of resident population; and error.

The independent regional registries differ considerably in their methods of data collection; some employ peripatetic clerks, others use hospital record staff to extract data for the registry, and several rely heavily on other organisations' computer systems including those in hospitals and pathology laboratories. The registries probably also differ in the **completeness** of their data, but the best are known to have very high levels of completeness. Direct measures are only available from occasional special studies.[3,4] That by Hawkins and Swerdlow[3] estimated that the incompleteness of registration of childhood cancers by the regional registries was just under 5%; incompleteness may be greater for adults, for whom registration and record linkage (in the registries and at NHSCR) may be more difficult, than for children. General indications of completeness can be obtained from comparisons of the numbers of registrations and deaths in a period. The figures for deaths are those coded to a particular type of cancer as the underlying cause of death in residents of the same geographical area. Such mortality to incidence ratios by sex and site for 1992 are presented in **Table 9**. These ratios have several limitations, but there are variations between regions (and over time) which would be difficult to explain unless there were similar variations in completeness.

As with completeness, the **accuracy** of the data is only occasionally known directly from special studies. Various indirect measures, however, suggest that there is considerable variation between regions.

In addition, the **completeness of flagging** of registrations by NHSCR is important for cohort studies. The proportion of cancer registrations received by ONS which were successfully linked to an NHSCR record appears to have been about 96 per cent in recent years. The importance for any particular study of the records not traced will depend upon any biases by region, site or other main factors of interest.

Over the years, changes have occurred to the number of registries and to their **geographic coverage**. In 1950 there were 74 centres registering cancer in England and Wales, but the system was progressively simplified and by 1958 ten regions were covered by regional cancer registries; full coverage of England and Wales was achieved in 1962. Some registries covered more than one RHA: the current Thames Registry was formed in 1985 with the merger of the North West, North East and South Thames registries (the last covered both the South West and South East Thames RHAs). Wessex was separated from the South Thames registry in 1973; this coincided with a change in the method of data collection and a substantial increase in numbers of registrations for some parts of the Wessex region. Following recent reorganisations at the regional level in the NHS, the former South Western and Wessex RHAs are now covered by the South and West Cancer Intelligence Unit based in Bristol and Winchester. The former Yorkshire RHA and part of the former Northern RHA are now covered by the Northern and Yorkshire Cancer Registry and Information Service based in Leeds (the remainder of the former Northern RHA, South Cumbria, is now covered by the North Western registry). Some registries received reports from several centres in their region - at various times five regional centres existed in Trent, two in South Western, and three in East Anglian.

The point in time at which ONS, in consultation with the regional registries, decides to produce the tables for the reference volume is necessarily a compromise between two principal considerations - the need to minimise the delay between the relevant data year and the publication of the detailed results, and the requirement to obtain a very high level of completeness of the data and hence minimise the number of **late registrations.** The gap between the data year and production of tables has varied considerably; as a result there are currently (July 1998) varying proportions of additional cancer registrations held on the computer files at ONS compared with the numbers published in the corresponding reference volume, as shown in Figure 1A. Up to 1985 the annual differences range from 0% to 12%. Over the nineteen year period the differences has averaged around 4% although the differences for 1985, 1986 and 1987 are larger as a result of the problems with the transmission of data between the Thames registry and ONS.[5] The overall figures contain within them some substantial variations among the regions. For example, a problem at ONS with the processing of one data tape for 1985 from the North Western registry resulted in a shortfall in the published figures of around two thousand registrations. Although this made a difference of less than 1 per cent to the total for England and Wales, it represented a shortfall of around 10 per cent for the North Western region. **Late deletions and amendments** to data are in general a much smaller problem than late new registrations.

While late registrations result in the figures published in the reference volume being too low, **duplicate registrations** can artificially inflate them. Such duplication may arise if a patient is resident in one region but treated in another; this is particularly so for those resident in North Wales and treated in Liverpool, and for those resident around London who are treated in central London. Duplications are prevented firstly by the regional registries which hold alphabetic indexes of names and carry out computer searches; and secondly by the flagging at NHSCR, where if on flagging, a previous registration is found for the individual, the registrations are examined to see if they are duplicates or true multiple primary cancers. The rules for decisions on duplicates/multiples have changed over time, particularly for 1978 registrations which led to a 13 per cent decrease in registrations for Welsh residents. Currently, with the agreement of the regional registries, all such cases are referred back to them by ONS, and decisions taken according to an agreed set of rules.

Since the early 1960s, copies of all **death certificates** mentioning cancer have been sent by ONS to the registry covering the RHA in which the death occurred. Any cancers registered solely from the information on the death certificates were not included in the published information prior to 1974, at which point an abrupt increase occurred. Registries use the death certificate information in different ways. For example, some check the data by reference to clinical notes or other local data sources, but others simply enter the death as a registration (with the year of death as the anniversary year).

Inaccuracies and incompleteness may arise from **diagnostic practice,** and changes in it, although such errors and changes come from outside the registration system and are not under its control. Misclassification of cancers is more likely to occur when there is no opportunity to obtain histological confirmation of disease, or if the tumour has a pre-malignant stage which can be confused with invasive carcinoma. Misclassification may also result from mistakes in the collection, abstraction or coding of information both before and after it reaches the registry. Also, **clinical and pathological** (and registry) **definitions** of cancer may change over time and between places, particularly for borderline malignant conditions.

Changes in **coding systems** may cause discontinuities in published data. For the national data held by ONS, from 1971 to 1978 site is coded to ICD8 and histology by the Manual of Tumor Nomenclature and Coding (MOTNAC) 1968 edition; from 1979 onwards, site is coded to ICD9 and histology to ICD-O. Details of the effect of the changes between the ICD revisions on mortality statistics have been published;[6] these give an indication of their likely effect on cancer registrations. In addition, there have been some minor changes in ONS coding and classification rules.[3]

Over the past 25 years, submission of data from the registries to ONS on abstract cards has been superseded by computer media (punched cards, magnetic tape and diskettes). Abstract cards were coded at ONS whereas magnetic tapes and diskettes were coded by the registry before being sent to ONS. Thus a change to magnetic tape (the last region to do so was Oxford in 1985) may have been accompanied by changes in interpretation of coding.

Figure 1A Number of registrations (thousands) published in ARVs and currently on database

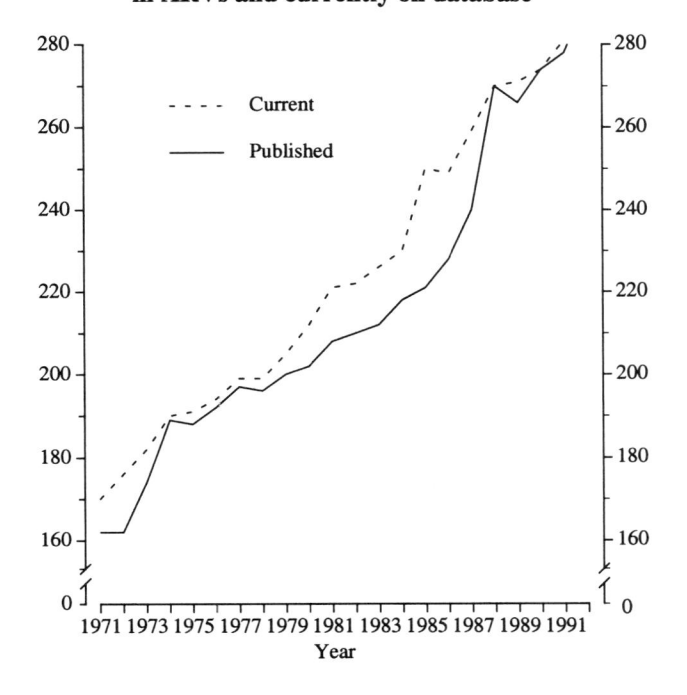

Rates of cancer incidence are dependent not only on the accuracy of the cancer registration data but also on that of the **population** denominator data. Recent censuses are believed to have been very accurate overall: under-enumeration in 1981 was estimated to be 0.5 per cent (240,000 people) and in 1991 to be 1.1 per cent (572,000 people), but this varied by age and by geographic area. Annual mid-year estimates of population, based on census data together with information on births, deaths and migration (see above) also appear to be very accurate on a national basis, although errors of several per cent have been found for some counties, districts and London boroughs. There may also be differences between the definitions of 'place of residence' used for cancer registrations and for population estimates. For the former, the address used is 'the usual place of residence as given by the patient', whereas the census definition is not so straightforward, particularly when a person lives at more than one address throughout the year.[7] This may lead to biases in analyses of data for small areas which include large numbers of students, armed forces or people living in institutions.

Finally, in published data on the scale of the national cancer registration system it is almost inevitable that straightforward **errors** will occur, for example in the transcription and printing of tables. Corrections to known errors have been published. The vast majority of the tables in this volume are computer-generated and transformed into camera-ready copy, which considerably reduces the potential for error.

Standardised registration ratio (SRR)

The incidence of cancer varies greatly with age. The SRR is an index which enables ready comparison of incidence rates in populations with different age structures. It is calculated by denoting one set of age-specific rates as the standard. These are then applied to each of several index populations of known age structure to show how many registrations would have been expected in these index populations had they, at each age, experienced the cancer incidence of the standard population. The 'expected' incidence so found is then compared with the observed, their ratio being multiplied by 100 to give an index in which 100 is the value for the standard population.

Calculations for 1992 are based on nineteen age groups (those used in Table 1).

The use of SRRs enables data for a particular site and sex to be presented as a single index figure relative to a defined standard or baseline. If the incidence patterns in the various age groups are different in the two populations or time periods, however, SRRs are an unreliable guide to comparison, and age-specific rates should be examined.

Table 6 shows the 1992 SRRs in RHAs of residence. For each cancer, the 1992 registration rates in England and Wales are taken as the standards (with the sexes considered separately).

For example, the SRR for cancer of the stomach in the Northern RHA was calculated as

$$SRR = \frac{100 \times \text{No. of registrations of cancer of the stomach in Northern RHA}}{\sum_{\text{Age group}} \left[\begin{array}{l} \text{Population in each age group, Northern} \\ \text{RHA} \times \text{registration rate for cancer of the} \\ \text{stomach for that age, England and Wales} \end{array} \right]}$$

Populations

The population figures for 1992 used to calculate incidence rates given in this volume are the mid-year estimates of the population resident in England and Wales, based on the 1991 census. The mid-1990 estimates are not directly comparable with those produced for years before 1981: residents who were outside Great Britain on census night are now included whereas overseas visitors to Great Britain are now excluded.

Table 2 shows the populations for each regional health authority, by sex and age. Users requiring further information on these estimates should contact the Population Estimates Unit at:

Office for National Statistics
Segensworth Road
Titchfield
Fareham, Hants
PO15 5RR

Occasional Paper No. 37 describes methods used by ONS to produce annual mid-year estimates of the population of local and health authority areas in England and Wales. It includes historical background and methods used in the 1980s. Details are given of the components of change (births, deaths and migration), and of methods used to estimate some special groups in the population, such as students and armed forces. Methods for rebasing the estimates for the 1990s, incorporating the results of the 1991 Census, are also included. The paper is available, price £4.00, from ONS Sales Desk at:

Office for National Statistics
B1/04
1 Drummond Gate
London SW1V 2QQ

Regional health authorities

Cancer registrations are recorded by cancer registries based on the registry or regional health authority area. The tables in this report are presented by the administrative RHA of usual residence.

Some regional cancer registry publications present regional statistics based on the number of patients treated in the cancer registry area. Therefore statistics in some regional cancer registry reports may differ from the region of residence analyses shown in this report.

From 1 April 1982, a single structure of 192 district health authorities replaced the former area health authorities and health districts in England. Certain regional health authorities were also affected by boundary changes. As a result of these and subsequent changes, the 1992 RHAs as defined in terms of the new district health authorities were as shown in Appendix 3 (on page 72).

Survival

ONS registrations since 1971 have been linked at the NHSCR to the death records (as already described); the latest available national survival tables were published in *Cancer survival in England and Wales: 1981 and 1989 registrations*.[11] Extensive material for England and Wales is given in *Trends in cancer survival in Great Britain*.[12]

The results of the EUROCARE cancer survival study, which covered 30 cancer registries in 12 European countries, were published[13] in May 1995. Six cancer registries in England participated; these were geographically spread around the country and covered almost half the population. Cancer registration data up to 1985 were included.

As noted above, the work at ONS on the redevelopment of the cancer registration processing system has now been completed, considerable improvements have been made to the quality of the records on the database, and all deaths have been linked to registrations for incidence years up to 1992 at the NHSCR.

Symbols and conventions used

-	nil
..	not available
:	not appropriate
nos	not otherwise specified
nec	not elsewhere classified

Further information

Special tabulations are available to order (subject to confidentiality thresholds) and on repayment. Such requests or enquiries should be made to:

National Cancer Registration Bureau
Office for National Statistics
B6/02
1 Drummond Gate
London SW1V 2QQ

References

1. WHO. *International Classification of Diseases, Injuries and Causes of Death, Ninth Revision*. Geneva (1977).
2. Swerdlow A J. Cancer Registration in England and Wales: Some Aspects Relevant to Interpretation of the Data. *Journal of the Royal Statistical Society* Series A, 1986, **149**, 146-160.
3. Hawkins M M and Swerdlow A J. Completeness of cancer and death follow-up obtained through the National Health Service Central Register for England and Wales. *British Journal of Cancer*, 1992, **66**, 408-413.
4. Villard-Mackintosh L, Coleman M P and Vessey M P. The completeness of cancer registration in England: an assessment from Oxford FPA study. *British Journal of Cancer*, 1988, **58**, 507-511.
5. OPCS. *Cancer statistics - registrations 1988*. Series MB1 no.21. HMSO (1994)
6. OPCS. *Mortality statistics - comparison of 8th and 9th Revision of the International Classification of Diseases*. Series DH1 no. 10. HMSO (1983).
7. OPCS and GRO(S). 1991 Census: Definitions, Great Britain. HMSO (1992).
8. OPCS. *Cancer registration surveillance, 1968-1978*. OPCS (1983).
9. Webber R and Craig J. Which local authorities are alike? *Population Trends 5*. HMSO (1976).
10. Craig J. Urban and rural local authorities. *Population Trends 8*. HMSO (1977).
11. ONS/ICRF. *Cancer survival in England and Wales: 1981 and 1989 registrations*, Monitor MB1 98/1.(1998).
12. *Trends in Cancer Survival in Great Britain*. Cancer Research Campaign (1982).
13. Berrino F, Sant M, Verdecchia A, Capocaccia R, Hakulinen T, Esteve J. (eds). *Survival of cancer patients in Europe. The EUROCARE Study*. Lyons: International Agency for Research on Cancer, IARC Scientific Publication No. 132 (1995).

Appendix 2 Cancer registries in the United Kingdom: current directors, addresses, telephone and fax numbers

Figure 2A Areas covered by the regional cancer registries, England and Wales, 1992

(a) England and Wales

Northern & Yorkshire	Professor R Haward, Medical Director
	Professor D Forman, Director of Information and Research
	Northern and Yorkshire Cancer Registry and Information Service, Arthington House Cookridge Hospital LEEDS LS16 6QB
	Tel: 0113-392 4163/4309 Fax: 0113-392 4178 df@yco.leeds.ac.uk

Trent
Dr J Botha, Director

Trent Cancer Registry
Floor 6
Weston Park Hospital NHS Trust
Whitham Road
SHEFFIELD
S10 2SJ

Tel: 0114-226 5351
Fax: 0114-226 5501
director@trentcancer.preslel.co.uk

Region	Contact
East Anglian	Dr C H Brown, Medical Director East Anglian Cancer Registry Addenbrooke's Hospital Hills Road CAMBRIDGE, CB2 2QQ Tel: 01223 216644 Fax: 01223 245636 twd10@medschl.cam.ac.uk Dr T W Davies, General Director Institute of Public Health Forvie Site Robinson Way CAMBRIDGE, CB2 2SR Tel: 01223 330318 Fax: 01223 330330
Thames	Ms E Davis, General Manager Thames Cancer Registry 1st Floor Capital House Weston Street LONDON, SE1 3QD Tel: 0171-378 7688 Fax: 0171-378 9510 s.groom@umds.ac.uk
Oxford	Dr M Roche, Medical Director Oxford Cancer Intelligence Unit Oxfordshire Health Old Road Headington OXFORD, OX3 7LF Tel: 01865 226742 Fax: 01865 226809 ociu@cix.compulink.co.uk
South & West	Dr J Smith, Director South and West Cancer Intelligence Unit Grosvenor House 149 Whiteladies Road BRISTOL, BS8 2RA Tel: 0117-970 6474 Fax: 0117-970 6481 South and West Cancer Intelligence Unit Highcroft Romsey Road WINCHESTER, SO22 5DH Tel: 01962 863511 Fax: 01962 878360 Jenifer@ciu.clra.net
West Midlands	Dr G Lawrence, Director West Midlands Cancer Intelligence Unit The Public Health Building University of Birmingham Edgbaston BIRMINGHAM, B15 2TT Tel: 0121-414 7711 Fax: 0121-414 7712 wmciu@wmciu.thenhs.com
Merseyside & Cheshire	Dr E M I Williams, Director Merseyside & Cheshire Cancer Registry 2nd Floor Muspratt Building The University of Liverpool LIVERPOOL, L69 3BX Tel: 0151-794 5690 0151-794 5691 Registry Fax: 0151-794 5700 emiw@iv.ac.uk
North Western	Professor C B J Woodman, Director North Western Cancer Registry Centre for Cancer Epidemiology Christie Hospital NHS Trust Kinnaird Road Withington MANCHESTER, M20 9QL Tel: 0161-446 3575 Fax: 0161-446 3578 cbjw@man.ac.uk
Wales	Dr J Steward, Medical Director Wales Cancer Intelligence and Surveillance Unit 14 Cathedral Road CARDIFF, CF1 9JL Tel: 01222 373500 Fax: 01222 373511 John.steward@velindre.tr.wales.nhs.uk

(b) Scotland

ISD Scotland	Dr D Brewster, Director of Cancer Registration National Health Service in Scotland Information and Statistics Division Trinity Park House South Trinity Road EDINBURGH, EH5 2SQ Tel: 0131-551 8903 Fax: 0131-551 1392

(c) N Ireland

N. Ireland	Dr A Gavin, Director Northern Ireland Cancer Registry Dept of Epidemiology & Public Health The Queen's University of Belfast Mulhouse Building Institute of Clinical Science Grosvenor Road BELFAST, BT12 6BT Tel: 01232 263136 Fax: 01232 248017

Appendix 3 Regional and district health authorities (1992)

Northern
Hartlepool
North Tees
South Tees
East Cumbria
South Cumbria

West Cumbria
Darlington
South West Durham

Northumberland
Gateshead
Newcastle
North Tyneside
South Tyneside
Sunderland
North Durham

Yorkshire
Hull
East Yorkshire
Grimsby
Scunthorpe
Northallerton

York
Scarborough
Harrogate
Bradford
Airedale

Calderdale
Huddersfield
Dewsbury
Leeds
Wakefield
Pontefract

Trent
North Derbyshire
Southern Derbyshire
Leicestershire
North Lincolnshire
South Lincolnshire

Nottingham
Barnsley

Doncaster
Rotherham
Sheffield
North Nottinghamshire

East Anglian
Cambridge
West Suffolk
East Suffolk
North West Anglia
Norwich

Great Yarmouth and Waveney
Huntingdon

North West Thames
North Bedfordshire
South Bedfordshire
North West Hertfordshire

South West Hertfordshire
Barnet
Harrow
Hillingdon
Hounslow and Spelthorne

Ealing
Parkside
Riverside
East and North Hertfordshire

North East Thames
Basildon and Thurrock
Mid Essex
North East Essex
West Essex
Southend

Barking, Havering and Brentwood
Hampstead
Bloomsbury and Islington
City and Hackney

Newham
Tower Hamlets
Enfield
Haringey
Redbridge
Waltham Forest

South East Thames
Brighton
Eastbourne
Hastings
South East Kent
Canterbury and Thanet

Dartford and Gravesham
Maidstone
Medway
Tunbridge Wells
Bexley

Greenwich
Bromley
West Lambeth
Camberwell
Lewisham and North Southwark

South West Thames
North West Surrey
West Surrey and North East Hampshire
South West Surrey
Mid Surrey
East Surrey

Chichester
Mid Downs
Worthing
Croydon

South West Thames - *continued*

Kingston and Esher
Richmond, Twickenham and Roehampton
Wandsworth
Merton and Sutton

Wessex
Dorset
Portsmouth and South East Hampshire
Southampton and South West Hampshire
Winchester

Basingstoke and North Hampshire
Salisbury
Swindon
Bath
Isle of Wight

Oxford
East Berkshire
West Berkshire
Aylesbury Vale
Wycombe
Milton Keynes

Kettering
Northampton
Oxfordshire

South Western
Bristol and District
Cornwall and Isles of Scilly
Exeter

North Devon
Plymouth
Torbay
Gloucester
Somerset

West Midlands
Bromsgrove and Redditch
Herefordshire
Kidderminster and District
Worcester and District
Shropshire

Mid Staffordshire
North Staffordshire
South East Staffordshire
South Warwickshire

West Midlands - *continued*

East Birmingham
North Birmingham
South Birmingham
West Birmingham

Coventry
Dudley
Sandwell
Solihull
Walsall
Wolverhampton
North East Warwickshire

Mersey
Chester
Crewe
Halton
Macclesfield
Warrington

Liverpool
St Helens and Knowsley
Southport and Formby
South Sefton
Wirral

North Western
Lancaster
Blackpool, Wyre and Fylde
Preston
Blackburn, Hyndburn and Ribble Valley
Burnley, Pendle and Rossendale

West Lancashire
Chorley and South Ribble
Bolton
Bury
North Manchester

Central Manchester
South Manchester
Oldham
Rochdale
Salford

Stockport
Tameside and Glossop
Trafford
Wigan

In 1992 the district health authorities in Wales were as follows:

Clwyd
East Dyfed
Pembrokeshire
Gwent
Gwynedd

Mid Glamorgan
Powys
South Glamorgan
West Glamorgan